The Call of the Hen
The Science of the Selection and Breeding Poultry For Egg Production

by Walter Hogan

with an introduction by Jackson Chambers

This work contains material that was originally published in 1914.

This publication is within the Public Domain.

*This edition is reprinted for educational purposes
and in accordance with all applicable Federal Laws.*

Introduction Copyright 2017 by Jackson Chambers

Self Reliance Books

Get more historic titles on animal and stock breeding, gardening and old fashioned skills by visiting us at:

http://selfreliancebooks.blogspot.com/

Introduction

I am pleased to present yet another title on Poultry.

The work is in the Public Domain and is re-printed here in accordance with Federal Laws.

As with all reprinted books of this age that are intended to perfectly reproduce the original edition, considerable pains and effort had to be undertaken to correct fading and sometimes outright damage to existing proofs of this title. At times, this task is quite monumental, requiring an almost total "rebuilding" of some pages from digital proofs of multiple copies. Despite this, imperfections still sometimes exist in the final proof and may detract from the visual appearance of the text.

I hope you enjoy reading this book as much as I enjoyed making it available to readers again.

Jackson Chambers

93350B

DEDICATED
TO THE POULTRYMEN WHO,
LIKE THE AUTHOR,
DO NOT KNOW IT ALL.

Lady Show You, a White Plymouth Rock hen, that holds the world's egg record for a two-year-old hen; laid 281 eggs in the National Egg-laying Contest at the Missouri State Poultry Experiment Station, Mountain Grove, Mo. She met the Hogan test.

Photographed by request of the Chamber of Commerce, Petaluma, Calif.

These hens weighed less than 4 pounds each and laid 131 pounds 2 ounces of eggs. They won the prize for laying the greatest weight in eggs in the National Egg-laying Contest. Each hen's eggs would have sold for **$4.50 on the Petaluma Market**, if reduced to No. 1 eggs. They are the result of five years' breeding by the author from common Petaluma Single Comb White Leghorns. It is possible for the reader to do the same with almost any breed by following instructions in this book.

PREFACE

This is an age which demands action, applied thought, and a practical, actual, and workable science. The world is demanding to know, not "What are you?" or "What do you look like?" but "What can you do?" Drones are being culled out in all lines of business activity, and rightly so; and the same is true with the poultry business. The hen which delivers the goods is the hen which is in demand. "The hen that lays is the hen that pays."

We have two reasons for publishing THE CALL OF THE HEN. Some three years ago Mr. Hogan sent us three males, all Single Comb White Leghorns; one was of his 280-egg type, selected according to this system, another was of the 250-egg type, and the third was of a 70- or 80-egg type. He also sent us two pens of hens of his own selection and breeding. We trapnested all the hens, and bred from all three males. The results in every case have borne out Mr. Hogan's claims and the truthfulness of his methods of selection and breeding. We have also tested the hens in the egg-laying contests; taken measurements and made tests and judged their capacity for laying as per this system, THE CALL OF THE HEN. The results so nearly tally with the system in practically every case that we feel that this is a valuable method of selection and breeding, which should be in the hands of everyone who attempts to raise poultry.

Capacity, condition, type, and vigor must all be taken into consideration in determining whether a hen will be a good producer or a poor producer. By making a careful and sensible application of the rules made known in this book, it is possible for any poultry-raiser to avoid great loss.

We are told, and have good reason to believe that it is true, that the average farm hen lays less than 80 eggs per year. If that be true, about half the poultry is being kept at a loss to the owner. If this is the condition, are we not justified in doing something to attract the attention of the farmers and poultry-raisers to methods and practices which will lead to the production of more eggs from the average hen, and to the necessity of culling and selection, and to more careful and painstaking methods?

The object of THE CALL OF THE HEN is to stimulate an interest in increased egg-production in all varieties of poultry and to encourage the breeding of strains of high-producers. We have come to the point where our efforts to breed fowls with perfect plumage for show purposes has overshadowed that of the ability of our hens to lay; and it can certainly result in no harm to call the attention of the breeders of the nation to the good which would certainly come from a study of the things which would tend to increase egg-production. We should all be vitally concerned in any attempt to better conditions, to increase the productiveness of the hen, and to give impetus to an industry which is already one of our greatest agricultural factors.

For half a century the fanciers and poultrymen generally have devoted their attention to the show-room in the development of shape and color. No opportunity has been offered or anything specially done to encourage the farmer and poultryman to develop the natural resources of the hen—her ability to lay eggs. A few of our best experiment stations have made some investigations along this line and done some very valuable work indeed. Here and there an occasional poultry-breeder has given some thought and attention to breeding for egg-production; but certainly, as a whole, the attention of breeders generally has not been along this line, and it seems that this important matter has been too much neglected.

Haphazard methods of mating and breeding don't pay, and indiscriminate methods cannot prove successful in building up a flock of laying hens. There never was a time in the history of this country when poultry and eggs were in greater demand; the price at which poultry and eggs sell has increased much more in proportion than has the price of feeds necessary to produce these products; but because the industry is flourishing to-day more than ever before does not justify us in continuing indiscriminate or foolhardy methods. The opportunity is ours to insure greater profits, if we will but carefully and systematically solve the problem which is facing us: "How can we insure a reasonably high average egg-production?"

The interests of the fancier are served through the show-room. If a breeder enters birds in a show-room and is beaten, he tries to improve his flock and perfect it by introducing new blood or by improved methods of breeding and careful selection. If he wins, he tries to keep his stock in that high state of perfection. It is just as important, and even more so, that he know just what his flock can do in the matter of production, and he ought to use the same care in trying to perfect his strain of layers.

There are exceptions to all rules. You will find some exceptions in selecting, testing, and breeding your poultry according to the method described in THE CALL OF THE HEN; but many breeders have tested it for some six or eight years; many of these have doubled their egg-yield in this time. We feel certain that Mr. Hogan's method of selection and breeding will prove him to be to the poultry industry what Burbank is to horticulture, Edison is to the electrical world, or Darwin or Mendel to the breeding kingdom. That the mastery of this method of selection and breeding, and sensibly applying the principles revealed herein, will mean much to the poultry industry, is our honest belief.

AMERICAN SCHOOL OF POULTRY HUSBANDRY.

Mountain Grove, Mo.

FOREWORD

The writer's introduction into poultry-keeping was in the city of Boston, Massachusetts, in the autumn of 1857. By the the spring of '68 I had a flock of nearly 400 birds, among them a lot of the best Single Comb White Leghorns that I could find. I went in person to New York city to get them. My friends thought such extensive poultry-keeping the limit of folly, and freely remarked that I was going crazy. In those days eggs were almost worthless during the spring and summer months, but would often sell for fifty cents per dozen in the winter. This set me to thinking that perhaps it might be possible to increase the egg-yield in the winter and by so doing make the fad a better paying proposition. Through my experiments I found that all hens were not alike; that some would be very good table fowl and poor layers, others would be very good layers and poor table fowl, while still other hens would be very fair table fowl and very fair layers. At this time we had all the old-fashioned breeds we could get, and discarded them all for the Single Comb White and Brown Leghorns. I had decided that knowledge was of commercial value only when applied, and having a working knowledge of the anatomy and physiology of the hen, I decided to try to turn the same to a commercial account, and in a couple of years had evolved what is now known as the "Walter Hogan System," which consists in ascertaining the value of a hen for the **purpose you desire** by the relative thickness of and distance apart of the pelvic bones. Before 1873 I had communicated this discovery to some of my friends under promise of secrecy. One of them, Albert Brown, once a well-known banker, of Amesbury, Massachusetts, and O. H. Farrar, of the same place, an overseer in the Hamilton Mills, and a light Brahma specialist. After using the above so-called "system" for a number of years, I developed a new method, which I have taught in part privately for some years, and which I now introduce to the public under the title of "The Call of the Hen; or, The Science of Selecting and Breeding Poultry."

My friends early prophesied that my penchant for invention would land me in the poor-house in my old age. So by some occult inspiration I was induced to abstain from publishing any part of my discoveries until 1904, when, by the advice of Ex-Congressman Haldor E. Boen, of Minnesota, to whom I had confided my poultry secrets some years previous, I decided to publish only my first discovery, known as the "Walter Hogan System" (which will be found in the latter part of this work), after the same had been tested at the Minnesota State Experimental Station by Professor Hoverstadt, the superintendent of the station. However, before taking any steps to bring this matter before the public, I wrote to some thirty or more poultry judges, who were supposed to be selected as judges to officiate at the coming poultry show to be held in Buffalo during the exhibition at that place in 1901, asking them if they knew of any way to tell when a pullet was about to lay. I thought that if they did not know that much of the laying proposition, I would be safe in going ahead with publishing my secrets. The letters I received were left in Minnesota when I came to California shortly before the earthquake in 1906, so I cannot name the judges at present, but they will remember me as the proprietor of the Fergus Falls Woolen Mills; and I must say they replied in a very courteous manner, saying there was no way except the general appearance of the bird, as to its maturity of form, redness of comb and wattles, singing, looking for nest, etc. One only of the number charged me one dollar for this information.

Failing health obliged me to dispose of my manufacturing business and retire to the farm, and it was in the spring of 1905 before I published my "Walter Hogan System," when it appeared in a number of poultry papers. (See *Reliable Poultry Journal*, March, 1905.) I did not copyright the work at that time, although my experience in mechanical inventions had taught me that I should have done so, and the following August imitations began to appear, until in 1912 a number of different parties in the United States and foreign countries were claiming authorship and selling it under the same or different titles.

My years of research and expense brought me no financial returns, and in the spring of 1906 I left Minnesota for California, a physical and financial wreck. After having regained my health, I began here at Petaluma to build up the same kind of a flock of layers that I had done in previous years, with the idea of publishing my entire work when I should have bred up a strain of 200-egg hens and better.

After I removed to California, Professor M. E. Jaffa, of the University of California, became interested in the matter, and, at the request of the Petaluma Poultry Association, had the discovery tested at the California Poultry Experimental Station for two years, and continued for two years longer for the purpose of determining the value of four-year-old hens as layers, as it is outlined in this book in the chapter relating to the selection of the best layers in a flock.

It was also tested in New Zealand by D. D. Hyde, chief poultry expert for the New Zealand Government, and Prof. Brown, of the New Zealand Poultry Experiment Station. I have repeatedly been requested by my friends in different parts of the world to publish the full matter in book form, but poor health and lack of sufficient funds have prevented me from doing so until now. As this work will be copyrighted, I do not anticipate the literary pirates will raid it as they have my former work. In justice to the poultry fraternity, I want to say that while I have been and am now a member of the American Poultry Association, and have raised poultry fifty-six years, and now raise them by the thousand, I have never in the past classed myself as a "poultryman" in the strict sense of the word. Neither do I claim that I am the only one who has discovered the facts set forth in this book. I only know that I have never seen them in print before. I know what the results of following this method have been with me, and I feel safe in assuming that the things I have discovered have not been known. Hundreds have known me as an inventor and woolen manufacturer where one would know me as a "poultry crank"; and the only apology I have for offering this book to the public in a field already crowded with poultry literature is the earnest solicitation of my friends.

WALTER HOGAN.

Petaluma, Cal., July 7, 1912.

The Call of the Hen; or, The Science of the Selection and Breeding of Poultry

By WALTER HOGAN.

CHAPTER I.

The Underlying Principles Which Govern the Selection and Breeding of Poultry Are Capacity, Condition, Type, Constitutional Vigor, And Prepotency.

In the winter of 1910 I received a letter from a woman in Oregon, which read as follows:

"Dear Sir,—My husband is a machinist. He is getting old and his health is failing. We have both worked hard all our lives, and have saved enough to buy a small place in the country. We can no longer do hard work, and in looking for some light occupation that would bring weekly returns, we have looked favorably on the poultry business. We have kept a small flock of hens on a town lot for a number of years, and think we have done well with them. We also take four poultry papers, but each one tells a different story, and we cannot decide what to do. We have been years accumulating our little savings, and if we should lose them, we would have no resources left for our old age. I enclose two articles from the September (1910) number of the *Pacific Fanciers' Monthly*. One article gives me to understand that it is almost hopeless to think of making a living with hens, if we depend on selling eggs and poultry on the market. The other article holds out the promise of a possible income of a thousand dollars per year from 300 hens, if handled under right conditions. One means utter failure and bankruptcy in market eggs and poultry, and the other means the fullest measure of success. Both of these articles are in the same number and one follows the other on the same page. How can you reconcile these two conflicting opinions?"

(The articles follow.)

"A Common Question Wisely Answered.

"By George Scott.

"Can a living be made from poultry? Probably there is no one who has attained distinction in the avicultural arena to whom this question has not been put hundreds of times; and it is a question of perennial interest to the poultry-keeping public. There are many people who will tell you that a living, and a good living, can be made from poultry-keeping alone, and as proof of their statement will point out the numerous men whose names are household words in the fancy. On the other hand, a vast majority will most emphatically give utterance to statements calculated to deter any poultry-keeping aspirant, and give weight to their contention by citing hundreds of cases where men have tried and failed. Truly the mass of evidence appears to be with the latter belief, for it is an indubitable fact that for every person who succeeds in this business a hundred fail. But, looking at the matter from a logical point of view, the fact that a minority rely on poultry for their daily bread is ample evidence that it is quite possible to make a living out of poultry-keeping, and the abnormal number of failures merely proves that the business is a difficult one.

"The fact that a man who has failed in some other business takes up poultry-keeping with a like result in no sense proves that poultry-keeping does not pay; it is only what could be expected, and any experienced aviculturist would have prophesied such a result. It is, however, useless to explain such things to the man who is contemplating starting a poultry farm. To suggest that he is unfit for the task would be taken by him as an insult, for the public, in its ignorance, has conceived the idea that poultry-management is the simplest work that anyone can think of—in fact, I question whether an outsider considers it to be work at all.

"Such a hold has this belief obtained on the man in the street that it almost amounts to a superstition, and until the fallacy is exploded the number of the unsuccessful will be constantly increased. The public, apparently, cannot understand the difference between keeping a few fowls as a paying hobby and managing a poultry farm as an enormous one, and that the minor difficulties to be met with in the former case are increased a thousand fold in the latter.

"Probably there is no other business which calls for so many qualifications as that of the poultry-farmer, and to say that the man who has been successful in any other walk in life is totally unfitted for this business, though somewhat exaggerated, will give the tyro some idea of what is wanted. An intimate detailed knowledge of poultry-management, an unlimited reserve of perseverance, determination, and resource, a genuine love for fowls, the capacity for hard, continuous work for seven days a week, combined with business knowledge and thrifty management, are all essential, and will, with ordinary luck, lead one to the desired goal.

"I am very dubious as to whether a living can be made from utility poultry-keeping, pure and simple—that is to say, by selling eggs and birds solely for edible purposes. A profit can undoubtedly be made, but it is so infinitesimal that the income derived from this source alone would, I am afraid, scarcely suffice for the needs of the most parsimonious. If it is decided to specialize in utility points, pure-bred stock must be kept of the popular varieties, and eggs for hatching, day-old chicks, and stock birds must be sold. This will make all the difference, and once a connection has been worked up, there is no reason why the business should not pay, and pay well.

"The breeding of exhibition birds is, without doubt, the most profitable branch, and when once a name has been made, stock and eggs can be disposed of at most remunerative prices. Success, however, cannot be attained at once; it is often the work of years; and many breeders never rise from the ranks of mediocrity. Moreover, much capital is required to start an exhibition poultry farm, and one's expenses incurred in the management are infinitely heavier than is the case where utility points are the only consideration.

"I would not advise anyone unversed in poultry-culture to give up a situation, however poor, in order to go in for poultry-keeping as a means of earning a livelihood. To think of such a thing is foolish in the extreme, but for anyone to burn one's boats behind one in this way would be suicidal. What I would suggest to poultry-keeping aspirants (and I believe the number of these reaches well into four figures) is that they should keep as many fowls as they can attend to properly in their spare hours, and see what profits they can make from the birds. Above all, they must find out if they have a genuine love for the work, for without this nothing can be done. When a name has been made as a breeder of good stock, then, and then only, is it time for the amateur to consider the advisability of adopting poultry-keeping as a business; and long before this point is reached the glamor of the idea may have faded, for the life of a poultry-keeper is, contrary to popular belief, far from being a bed of roses. Practically all the men who are to-day making a living from poultry commenced keeping fowls as a hobby, and the knowledge and experience which they gained in this way enabled them to found the establishments which are to-day of world-wide reputation.

"To those who are qualified for the work poultry-keeping offers a good living; but to the idle, the thriftless or the pleasure-seekers of this holiday-making age it offers more desolate prospects than any other trade or profession. In this business nothing but dogged determination will enable the beginner to climb the rugged, precipitous path to success, and anyone who is lacking in this essential, or who is afraid of hard, continuous work, will save himself the obloquy of failure by choosing some other field in which to exercise his powers."

"The Good Little Hen.

"What She Will Do for You if You Will Treat Her Right.

"By Mrs. A. Basley.

"There is money in poultry for the man and especially for the woman that will dig it out. This I can assure the *Fanciers' Monthly* readers, if they are in doubt.

"'Dig it out' seems a curious way of putting it. When I spent a summer in a big mining camp in Colorado, I noticed a great many holes in the sides of the mountains. 'Yes,' said a miner, 'and not 5 per cent of those holes have paid.' It was appalling to think of the thousands of dollars lost in those holes. 'Give me a hundred hens,' said I. The money it took to dig one of those unprofitable holes would have started a fine poultry plant and the good little hens would have brought in a living for their owners.

"There is money in poultry! Every inch of a hen is valuable. I would like to give you one of the values in the hen and what it costs to keep her.

"First, there are the eggs she will lay, if properly fed and treated. Twelve dozen eggs per year is the average, although I personally know poultry plants now being operated in Southern California where the output, as shown by carefully kept records, is sixteen dozen per year. The average price at the Arlington Egg Ranch for the past year was 31 cents a dozen, because the proprietor arranged to have his hens laying when eggs cost the most, in the fall and winter months.

"Sixteen dozen eggs at 31 cents a dozen means each hen brings in $4.96 in eggs, whilst her food cost 10 cents per month or $1.00 per year, leaving $3.76 as profit for eggs.

"There is still another source of profit in the hen, and that is in the droppings. At several of the experiment stations it has been found that a hen voids about 100 pounds of droppings per year. These droppings have been analyzed and show a value as fertilizer of from 30 to 35 cents per hen; the value being controlled not only by the market demand, but also by the quality; the droppings being richer as fertilizer where the food was rich in protein and where the hens are fed the 'full and plenty' method.

"'What do you do with the hen droppings?' I asked a beginner. 'Throw them away; glad to get rid of them,' was the reply. At the rate of $10.00 per ton, that was a waste of 50 cents per hen. Two of our neighbors had lawns which were in so bad a condition from the soil being worn out that they were on the point of having them dug out and new soil put in and the whole re-sowed, when they thought of their hen droppings; these they had spread over the lawns and then raked off again and the lawns well watered. In a month's time those lawns looked beautiful—better far than if they had been re-made, and at far less cost.

"When I lived in the Eastern States, my window garden was the envy and admiration of everyone that passed; there were flowers galore all through the dark winter gloom and cold frosty days. I loved my plants, took good care of them in every way, but the secret of the wonderful blossoms was hen manure!

"Once a month I half-filled a bucket with hen droppings, poured a kettleful of boiling water on it, filling the bucket with the water, stirred it with a stick, let it settle and cool, and watered the plants with that liquid. I found that hen droppings enrich the ground for almost all plants better than anything; roses are the only exception that I have found, they doing much better when fertilized with well-rotted cow manure.

"But to return to our hen. She gives 30 pounds' weight of eggs, or sixteen dozen, valued at $4.96; she also gives 100 pounds of valuable fertilizer, worth here $10 a ton, or 50 cents per hen, which brings the amount of her earnings to $5.40, and at the end of the year we still have the hen to eat or sell at market value, about 75 cents or $1.00. If we eat her, we have the feathers, which are easily saved and can be sold or made into pillows, the bones pounded up and fed to the other fowls.

"Poultry pays, and pays better than any other legitimate business, considering the amount invested. Why then are there any failures? I will tell you why: The failures are not the fault of the good little hen. She will always do her duty; she will always respond to the treatment she gets. The failures are the people who care for the hen. The owners are the failures, and not the fowls.

"Success is what we all want to attain in whatever we undertake; and, 'lest we forget' some of the things which lead to success, may I repeat that there are three essentials to egg-production? These are: Comfort, Exercise, and Proper Food. I would like to review these."

I wrote the lady that both of these articles were right. Let us see if we can prove the statement. If the reader has ever had any experience with cattle, he knows it would be sheer folly to buy a herd of Polled Angus or Herefords for a dairy farm, for they have been bred for years for beef, and practically everything fed to them goes to meat; while it would be just as foolish to buy a herd of Jersey cows and expect to make a living from them raising beef, as they have been bred for years for butter-fat, and practically everything fed to them goes to milk and cream. If the reader's experience has been with horses, he is aware that a man engaged in teaming would not select the trotting type of horse, neither would a turfman put his money on an 1800-pound Clyde horse, if the balance of the field were trotting horses; that would not be horse sense. Now, the same comparison holds good in the poultry field, except with this difference, that the egg type and meat type in poultry have never been segregated into different breeds, and each breed bred for a number of years along the line it was intended for—the egg type bred for **eggs alone,** and all birds inclined to meat-production discarded—both **male** and **female,** and the meat type bred for **meat,** without regard to eggs, except **enough to perpetuate the species,** just as the typical butter cattle and typical beef cattle have been bred.

I have seen a great many cases like the first-mentioned article, where a person would go into the poultry business and get started with stock that was of the meat type, and, not knowing any better, would think that all poultry was the same as his, and the only way any money could be made in the business was to sell fancy birds and eggs at fancy prices. Now, these people are not to blame for what they do not know. They think their hens are as good layers as any other hens, and they have no way of knowing any better.

I have also seen a great many cases like Mrs. Basley writes of, except the profits were not so large, owing to different environment, I suppose. These people had the same **breed of hens** as the parties before mentioned, but they were fortunate in getting the **egg type,** and they made money with their hens. Everyone thinks every other person's hens are the same as theirs, if they are of the same breed, and that is the reason there are so many different conflicting statements in the poultry papers, and not because the writers are not intelligent or not truthful, as some suppose. From a scientific point of view, and apart from the fancy, and as far as the knowledge of meat- and egg-production is concerned, the poultry business **is in its infancy,** and the people who write for the poultry papers give their experience for your benefit. That is all.

To further impress on your mind the difference between poultry and other stock, I would say that while some individual cattle of the various beef breeds will not be a paying proposition, the only safe plan is to select your feeders from the beef family; and while some Jersey cows will not pay as butter-producers, still, as a breed, they are among the best for that purpose. Though some trotting horses do not make good, as a rule they will carry you over the road in good time, and though some draft-type teams are not sure pullers, they are a success as a class.

The same general laws apply to all animal nature. The hen is no exception, only in this respect: that while cattle and horses have been bred so that as a rule novices can select the type they wish by selecting the breed, hens have not been bred that way. We have what purport to be egg breeds and dual-purpose breeds. The first are supposed to be a paying proposition as a whole for egg-production. The latter are supposed to be a paying proposition for both eggs and meat combined; some breeders claiming that their breed will give you the very largest number of eggs per year and the greatest weight of flesh all in one bird. Now, these claims are misleading. It is an utter physical impossibility for any hen to be a typical egg type and at the same time be a typical meat type. It is against the laws of Nature. We have the Leghorns, Minorcas, Spanish, and a number of other Mediterranean breeds that are called "egg type." While the truth is, that while they have been bred as best the breeders knew how along the lines of egg-production, you can find vast numbers that will not lay eggs enough to pay for the feed they eat. Great numbers in some flocks have all the characteristics of the beef type, and will lay about three or four dozen eggs per year and sometimes not over a dozen. The Plymouth Rocks, Orpingtons, Wyandottes, and Langshans are classed as "dual-purpose" breeds, which means hens that will lay a medium number of eggs and give a good large carcass for the table; and while this is true in a majority of cases, I have seen numerous specimens that laid over two hundred and fifty eggs per year, while some would lay little or nothing. In fact, while I have bred Leghorns for more than forty years, and they are my favorite breed, I must say I have found as good layers (within a few eggs) in all the other breeds I have named as I have found in the Leghorns, and I have also found as poor layers among the Leghorns as I have found in any other breed. As far as the number of eggs is concerned, as a rule, I find that the breed of the hen has nothing to do with it whatever.

I do not wish to be considered dogmatic in anything I may say in this work. **I am merely giving the opinions I have formed by observation and experiment during a period of fifty-six years that I have kept poultry,** not to make all the money I could out of them, but to learn all I possibly could about them—in fact, until a few years ago I never kept poultry for the money there was in it. The keeping of hens has been a passion with me. I have spent years of time and thousands of dollars, but I think I have found something that will be of inestimable value to the world, and I have found it not because I was any better fitted for the work than thousands of other lovers of poultry, but because I stuck everlastingly to it, without any regard as to whether it paid me in dollars or not.

As previously stated, it is not a matter of breed as to whether a hen is a good layer or not. It is a matter of **type, capacity, and constitutional vigor.** First, in almost all breeds there is a type of hen where everything she consumes over bodily maintenance goes to the production of eggs. This we call the "typical egg type." Second, there is a type where about half the food consumed over maintenance goes to the production of eggs, the balance over bodily maintenance going to make flesh. This is called the "dual-purpose type," as this hen performs two functions that are considered necessary in the economy of Nature: the production of eggs and the production of meat on a commercial scale. Third, there is a type where everything consumed over bodily maintenance goes to flesh. This hen we call the "meat type," for the reason that practically all her energy is used in producing meat.

Now, here we have three distinct types of fowl in almost every breed. We have divided these three types into six separate classes for each type:

No. 1 of the **typical egg type** hen may lay about 36 eggs;
No. 2 may lay about 96 eggs;
No. 3 may lay about 180 eggs;
No. 4 may lay about 220 eggs;
No. 5 may lay about 250 eggs;
No. 6 may lay about 280 eggs.

All this is in their first laying year.

No. 1 of the **dual-purpose type** hen, may lay about 20 eggs;
No. 2 may lay about 50 eggs;
No. 3 may lay about 96 eggs;
No. 4 may lay about 115 eggs;
No. 5 may lay about 130 eggs;
No. 6 may lay about 145 eggs.

This in their first laying year.

No. 1 of the typical meat type may lay from nothing to a dozen eggs. Nos. 2, 3, 4, 5, and 6 may lay from nothing to a couple of dozen eggs, and, as a rule, will lay these in the spring when the crows lay. The reason is very plain, if we stop to think that the same natural laws govern all animal (and **human**) nature.

The egg type hen is of a nervous temperament (that is why she is usually free from body lice, if she has a suitable place to dust in), and all she eats over bodily maintenance goes to the production of eggs. The hen of the sanguine temperament is a little more beefy, and lays less eggs; the hen of the bilious temperament is more beefy still, and lays still less eggs, while the hen of the lymphatic temperament will lay little or nothing, almost everything she eats going to flesh and fat. (The reader need borrow no trouble over the meaning of the terms "nervous," "sanguine," "bilious," and "lymphatic" temperaments, if he is not familiar with them, as the charts 1, 2, 3, 4, 5, and 6 will specify matters so that anyone can understand the matter of selecting the different grades of hens with very little study and trouble.)

We have said that we have divided the three grades, the egg type, dual-purpose type, and meat type, into six separate classes. There is, in fact, a seventh class, but it is so rare that we will not take it into consideration here, but will explain it later. But we have, in fact, made ninety classes of these six for convenience in selection, and the process could be extended indefinitely, but it would serve no needful purpose.

Now, when we consider all these different grades in the hens of every breed, and the further fact that there is the same number of different grades in the male bird, is it any wonder that there is so much difference of opinion in regard to the profits derived from poultry-keeping? We have visited hundreds of poultry plants that numbered from about fifty to two thousand or more hens each. We have seen some flocks of five hundred that would not pay for the feed they consumed, for the simple reason that they were not the right type of hens. They were fine-looking, healthy meat-producers, but there was no earthly way possible to feed them that would induce them to lay eggs at any time except a few months in the spring when the crows laid, and eggs were cheap. The owners of some of these flocks were bright, brainy, vigorous business men, who tried every method that usage and science suggested, and fought with sheer desperation to make a success of the business, but went down in failure; while their next neighbor, a little pin-headed, conceited specimen of humanity, strutting around like a peacock, was getting rich with **the same breed of hens.** "Luck," do you say? Yes, it is mostly a matter of chance. The first man was unfortunate in that he got his eggs or breeding-pens from stock such as that described in the first article of the *Fanciers' Monthly*, while the last man got his eggs or breeding-pens from stock described by Mrs. Basley in the second article.

We once visited a gentleman who had a very extensive poultry plant. He had a large number of different breeds yarded off in finely appointed yards, with help and financial means to satisfy every need of a poultry plant. His pens of Rocks, Orpingtons, and Langshans were remarkable layers, while his Cochins, Houdans, and Polish were very good layers. After looking over the last-named birds, he remarked: "I have 500 Leghorn hens that are eighteen months old which I wish you would look at." After we had looked at them a few minutes, he said, "What do you think of them as layers?" I replied that if he would tell me which pen laid an average of all the pens, I would tell him in a few minutes. "That pen there," said he, pointing to No. 20, has laid an average number of all the eggs laid. I looked it up only last night." After examining the hens, I told him I would not take them as a gift, if I had to keep them one year. "Why?" he asked. "Because," I replied, "after keeping them a year and selling them, the price I would receive for the hens and the eggs they would lay would not pay for their feed. I cannot see why you keep them." The next evening he said to me, "Do you see that man moving into the place over yonder? Well, I have sold those Leghorn hens to that newcomer for $500." "Is this an exceptional case?" you ask. I have only this to say: that all the David Harums are not in the horse business, neither can I see why a poultryman should be his brother's keeper, **when it is not the rule in other lines of business.** It seems to me the better way is, to study poultry from a scientific point of view, so that you can judge the value of a hen for the purpose **you want her for,** and not have to depend on other people's opinions.

By studying this book carefully you will be able to tell approximately the number of eggs a hen is capable of laying in a year; you can also select the hens that will be the best for breeding purposes, for eggs, for meat, or as a dual-purpose hen—that is, a hen that will give you the largest number of eggs possible with the largest possible amount of meat when you wish to sell her, or the hen that will produce the best broilers, regardless of any one particular breed. Some hens will be very good layers, some very good meat-producers, some very good dual-purpose type, and some very fine fancy birds, and you can mate them with the same type of male bird and breed from these birds for a few generations, and **their progeny will degenerate.** The chickens from the hens and cockerels or cock birds of the 200-egg type may lay less each generation, until in eight or ten generations they may not lay enough to pay for their feed. The progeny from some of the best meat and dual-purpose type matings will sometimes degenerate just as the egg type, until they are practically worthless as profitable meat-producers. The chicks from the fancy mating may be a failure from the fancier's point of view.

This is the rock that some old poultry-breeders are sometimes wrecked upon. One case of national interest was the case of the late lamented Professor Gowell, of the State of Maine Experiment Station. He had started some years before to breed up a heavy-laying strain by using the trap-nest, selecting eggs for hatching from hens that were his best layers and conformed as near as possible to the standard, and using cockerels hatched from these eggs to mate with his hens. Now this was all right as far as it went, but there was something that the Professor had not taken into consideration. He had procured the best birds he could find, had trap-nested them to discover the hens that were the most prolific layers, had selected the eggs from what he had considered to be the best hens for the purpose (and few men had better judgment in this respect). He had mated up the best-looking cockerels from these best eggs from the best-laying hens, and according to all apparent precedents was he not justified in expecting an increase each year in egg-production? But what were the results? If reports are true, there was a decrease in egg-production, and what do you suppose was the cause? There must be some cause. There is a cause for every effect. Sometimes we think things just happen; that there is no natural law that governs them; that in this or that case it was all chance; that it may not have happened to another person, and will not be likely to happen to us again, and so we dismiss the matter only to have the same thing repeat itself, until we either solve the problem or meet our doom through it. And thereby hangs a tale.

Some time in the summer of 1905 I received a letter from a doctor in one of the suburbs of Boston, asking me what I would charge to visit Orono, Maine, and have a talk with Professor Gowell, and incidentally to drop a few remarks that might be of some help to him in his investigations. I had never met the Professor, but I replied to the Doctor that I would go (I was then living in Minnesota), and would pay my own expenses, as I wished to visit Boston, my birthplace, and where I first started in poultry-keeping in 1857, and it would be a small matter to go from there to Orono, Maine, where Professor Gowell was conducting his experiments. While I was waiting for a reply, I decided that as Professor Gowell had put so much time and thought into the trapnest proposition and had built so much on that one thing, and that as he could get results from it (only it was a waste of time), that in this first visit to him I would offer only one suggestion and that was the secret of selecting the birds, both male and female, that would be sure to breed progeny that would be better than their parents **along the lines in which the parents excelled,** or, in other words, transmit their predominating characteristics to their offspring; that is, if the cockerel or cock bird and hens were typical meat type birds, the progeny would excel along these lines. Some of them would excel their parents in the production of meat; they would be hardier, better feeders, would digest and assimilate their food better, and consequently arrive at maturity sooner, and be of **better flavor** and **more tender,** and by breeding these birds along the lines laid down by I. K. Felch, of Natick, Massachusetts ("line breeding" he calls it), they would improve each season, so that in a number of years there would be a great difference in their favor over their parents. If the pen was a fancy proposition and had been bred some years for fancy points, the progeny would show a decided improvement in a few years over their parents. If the pen were the typical egg type, the progeny would show an increase over their parents in **stamina** and **egg-production.** I would also have shown him where the birds he was breeding from were deficient in the faculty that governs fecundity, or, in other words, which controls the function of reproduction.

Whittier, in "Maud Muller," says, "For of all sad words of tongue or pen, the saddest are these: it might have been." Yes, "it might have been." Professor Gowell might have lived to give many more years of aid to the poultry world and his tragic death been prevented; but he wrote the Doctor tha the did not want me to come. He seemed determined to solve the problem himself, and no doubt would have done so if he had been as care-free from routine duties as a man in his position should have been; and I charge his untimely end to society. The men and women in our public institutions who are giving their lives for the benefit of humanity are not appreciated at their true value. We demand the full limit of routine duties, forgetting that it is impossible for a tired body to furnish sufficient nutriment to the brain to solve these intricate problems that are continually confronting them, and while we cause them to suffer mentally and physically individually, we cause ourselves to suffer collectively, by our parsimonious treatment of them.

CHAPTER II.

PRELIMINARY REMARKS, GIVING SOME ADVICE TO THE READER.

The writer is not one of the long-winded kind. I don't like to talk a long time in order to say a few words, or write a dozen pages where one will do as well. I believe in handing out the chunks of gold with as little dross as possible. I think the reader would rather receive the information I have to offer in one page than in a dozen; that he would rather discover the facts in a few feet than to be obliged to hunt over a hundred acres of literary space for the same information. For that reason I will make this work as brief as possible. I will be aided in my effort to do so by the fact that the theories offered in this work have been more or less demonstrated by the Government Experimental Stations of New Zealand and the States of Minnesota and California; also in the poultry plants of the five State hospitals (which contain thousands of hens) in the State of California, under the auspices of the State Board of Health and the physicians of the different hospitals. It might not be a difficult matter to mislead a few poultrymen on a subject that deals wholly with physiology and anatomy, but it would be absurd to think for a moment that one could deceive all the physicians in five State insane hospitals. It seems a man who would still doubt would believe the world is flat, especially when he learns that a member of the State Board of Health told the writer that there was a difference of $1,500 in favor of using the system, in one year, in one of the hospitals alone.

We commence in this chapter the unfolding of a method or test by which the reader can tell approximately the value of a hen and a male bird as a breeding proposition (and in the chapter on Breeding alone this book will be worth its weight in gold to the fanciers), an egg-producer or a meat-producer. It is my desire to make the facts contained in this book so clear and the tests so easy of application that anyone can become proficient in the use of them in a short time. Therefore I have prepared a series of illustrations showing numerous types and conditions of fowls, also various other facts that may better be shown by pictures than by explanations alone.

You will remember, no doubt, that you did not arrive at your present proficiency in reading in a day or two; that it took some little time, and there was a certain system or evolution in your study. You will find the same true of this method. There is a certain process that leads from one step to another, until you have covered the system, when by repeated study and practice you will become proficient and accomplish what at first seems impossible. It may seem an impossible task to handle and grade sixteen hundred hens in six hours, but the writer has done it. With sufficient help to hand me the hens, we graded (or, in other words, tested out) sixteen hundred hens in six hours in the State Hospital poultry yards at Ukiah, Mendocino County, California, in March, 1910. "Not so bad for a semi-invalid of 62," we hear you say. Our reply is, "It's practice." You can do the same. Go through the movements with every hen you pick up each day, and in a short time what at first is difficult will appear quite easy.

For some years previous to 1912 there was great activity in the poultry industry, there having been no lack of poultry papers, farm papers, and magazines that for a nominal sum would give tuition in poultry culture. The ease of getting a theoretical knowledge of the business induced thousands to take it up who otherwise would not have thought of doing so.

The apparent ease of conducting the business, the small amount of capital it was supposed to require, with the large and steady income it offered, were the will-o'-the-wisps that lured many to financial loss. I would warn my readers against rushing into the poultry business on a scale beyond their means without first obtaining a working knowledge of the same. With good stock, with the proper environment, a good market, and a working knowledge of the business, there is little danger of failure, if one is willing to do the work necessary on a poultry plant. It offers the most independent living for the smallest amount of capital of any business I know of.

The requisites for success are the knowledge of how to be able to select the hen you need for any particular purpose, whether it is for eggs or for meat or fancy; whether the hen will be a paying proposition or not (this may depend on your market); whether she will be able to transmit her predominating characteristics to her offspring or not. Also you must be able to judge accurately the value of the **male bird** as to what you want him for and as to his ability to stamp his offspring with the desired qualities. All the above you can learn from this book. You should also know how to operate incubators; how to feed and care for little chicks; how your hen-houses should be built to suit your climate; how your growing pullets should be fed and housed; and the best way to feed to get the most eggs at the smallest cost, and how to feed and mate to get fertile eggs and vigorous chicks. There are numerous books published on all of these latter subjects that you can buy from the publishers of any poultry paper; so we do not take up the matter in this work; we give only what you cannot get elsewhere.

Following is a series of half-tones and explanations representing the method we have used in instructing hundreds of poultrymen and women in California and other States and the managers of poultry plants in a number of State institutions in the State of California.

CHAPTER III.

THE VARIOUS STEPS IN THE APPLICATION OF THE METHOD OF SELECTION FOR EGG-PRODUCTION.

There are four characteristics that it is absolutely necessary for a hen to possess for the economical production of eggs or meat. The first is capacity, the second is condition the third is type, and the fourth is constitutional vigor. The reader must bear the first three in mind in studying the next few chapters, as we will dispose of these before taking other matters into consideration.

First. **What Is Capacity?**—Capacity means the abdominal capacity to consume and assimilate the amount of food necessary to produce the number of eggs or the amount of meat necessary to make the individual hen under consideration a paying proposition. We measure the capacity of the hen by placing the hand across the abdomen between the end of the breast-bone, or keel, and the pelvic bones. The method will be shown in detail in Chapter IV.

Second. **Condition.**—If the hen under consideration is an egg type, she must be kept in proper bodily condition by supplying her with the right quantity and quality of food that will furnish her with vitality to produce the number of eggs required of her. If the hen is in good condition, the flesh on the breast will be plump or practically flush with the breast-bone. Any variation in that condition will be shown by a shrinking away of the flesh of the breast, and will be followed by a corresponding shrinking of the abdomen. We show this by illustration and example later.

Third. **Type.**—She must be of a type that everything she consumes is used in producing the desired effect, whether it is meat, whether it is eggs, or whether it is the maximum amount of eggs and meat that a dual-purpose hen can produce. According to our idea, the type of hen determines how she will dispose of the food she eats. The kind of type is shown by the relative thickness of the pelvic bones. The very thin bone indicate the egg type. As we pass into the dual-purpose and beef types we find the bones becoming thicker. We show these by illustrations and charts later.

With the reader bearing the above three propositions in mind—namely, **Capacity, Condition, and Type**—we will proceed to show how to judge the hen with the least amount of time and labor.

Fig. 1—Showing hens in house.

Fig. 1 shows the interior of an open-front colony house, largely used around Petaluma. The roosts are connected to the house by hinges, so they can be hooked up out of the way while cleaning the house or examining the hens, as in the present case. These houses are usually about 8 feet wide and 10 feet deep inside, with 4 feet posts and pitch roof. These houses are open front, with the exception of 18 inches on each side, as can be seen on one side, where hens are going out of the house into the catching-coop. When hens move too slow to suit, one or more persons (children will do) can take a grain sack by bottom side in one hand and top side in the other hand and go into the house holding sacks spread apart and moving gently close to the floor or ground and drive the hens into the catching-coop. When the coop is full, shut down the slide door on outside to prevent hens returning to the house.

Some readers may have long houses, holding five hundred hens or more. In this case you will need a panel, run diagonally across the house to a point near the opening, where the hens go in and out of the house, as in Fig. 1½. This panel can be

Fig. 1½—Showing 2-inch wire panel placed diagonally across house holding 2,000 hens.

Fig. 3—Showing how hens are taken out of catching-crate.

as long as required for the width of the house and made in sections, if desired, and should be 6 feet or more high.

Fig. 2 shows hens in the coop. When there are enough in, we shut down the slide door and proceed as in Fig. 3.

Fig. 3. Note the slide door on top of the crate. We open this just enough to admit our arm while we grasp the hen firmly by both legs, so she can't twist around and injure herself. A slide door is better than a hinged door, as you can open the former just enough to take out the hen without so much danger of any of the other hens escaping.

Fig. 4. Note how the right arm is held in Fig. 4. This is not the right way, but it is the way most persons hold the left arm when they receive their first lesson. Now, note how the left arm is held; this is the right position, and it is difficult for me to teach students to hold their arms this way. I have to drill them repeatedly before they will do so. The hand which holds the hen by the legs should be at the height of the hip; this enables you to use the other hand in examining the hen for capacity with greater speed and accuracy.

Fig. 5 shows how the writer holds a bird to ascertain its capacity by holding it this way. After long practice, he is enabled to inspect one in a few seconds by having three parties to hand him the birds and to take them from him. **A small, light hen or pullet is best to practice with.**

Fig. 4—Showing right and wrong way to hold arms.

Fig. 5—Showing how a hen may be held while testing capacity.

Fig. 6—Showing where the hen's head should be so she cannot see anything.

Fig. 6 shows where the head of the bird should be. You will note that her eyes are covered up so she can't see, and that has a tendency to keep her quiet while you examine her.

Fig. 7 gives an example of testing the capacity of a hen. The hand is placed on the abdomen between the two pelvic bones and the rear of the breast-bone, the left hand holding the legs is turned under enough to bring the thighs away from the point of the breast-bone, so that the thighs will not interfere with measuring the depth of the abdomen. The depth of the abdomen will vary with different hens; some will be one finger (a finger means the width of a finger the widest way; I have called it three-fourths of an inch) between the two pelvic bones (sometimes called "lay" bones or "vent" bones) and the rear of the breast-bone. Some hens will be two fingers between the two pelvic bones and the rear of the breast-bone, some will be three fingers, some will be four fingers, some will be five fingers, some will be six fingers, and occasionally one will be seven fingers between the two pelvic bones and the rear of the breast-bone. The depth of the abdomen indicates the capacity or the ability of the bird to consume and assimilate food, and it applies to all breeds, except that, everything else being equal, the longer-bodied hen, having more

Fig. 7—Showing how to test capacity.

room for the digestive machinery, would have some advantage over the shorter-bodied hen.

Fig. 8. This indicates how to hold a hen when you examine her for condition. This is one of the most difficult and serious problems a poultryman has to deal with. To illustrate, I will cite one case out of hundreds that have come under my observation. A gentleman wrote me to call on him, as he was having trouble with his hens. When I arrived at his place, he told me that when he fed his hens well he got lots of eggs, but some of his hens died; then when he did not feed them so well they did not lay so many eggs, but none of them died. He said he had repeated this a number of times with the same results. He said the ones that died were as fat as butter. I picked up one of the hens; she was in prime condition for the market. I picked up another one; she was very thin. I examined all his hens. I found he had, like a great many poultrymen, three distinct types of hens: the egg type, the dual-purpose type, and the meat type. As he had fancy birds in all

34 THE CALL OF THE HEN.

the different types, he did not want to dispose of any of his flock, so I segregated them into three divisions: the egg type, the dual-purpose type, and the meat type. After that he fed the egg type all the grain they could clean up in the scratching-shed and kept a balance-ration of dry ground feed before them all the time. The dual-purpose hens were fed all the grain they could clean up in the scratching-shed, with a small amount of dry ground feed each day. The meat type hens were fed

FIG. 8—Showing how to test condition. The legs of the hen are drawn upward, so that you can see the breast. The condition is tested by placing the thumb and forefinger about ½ inch from the front of the breast-bone. Figs. 20, 21, and 22 show the method in detail.

a smaller amount of grain in the scratching-shed, with a couple of feeds each week of dry ground mash—just enough to keep them in condition. After this he had no more trouble with his hens not laying in the proper season and dying from being too fat. He would occasionally pick up hens in the different pens and note their condition and feed them accordingly. He told me later that before he had taken the lessons he had been working completely in the dark, but now he understood the matter thoroughly and knew what to do.

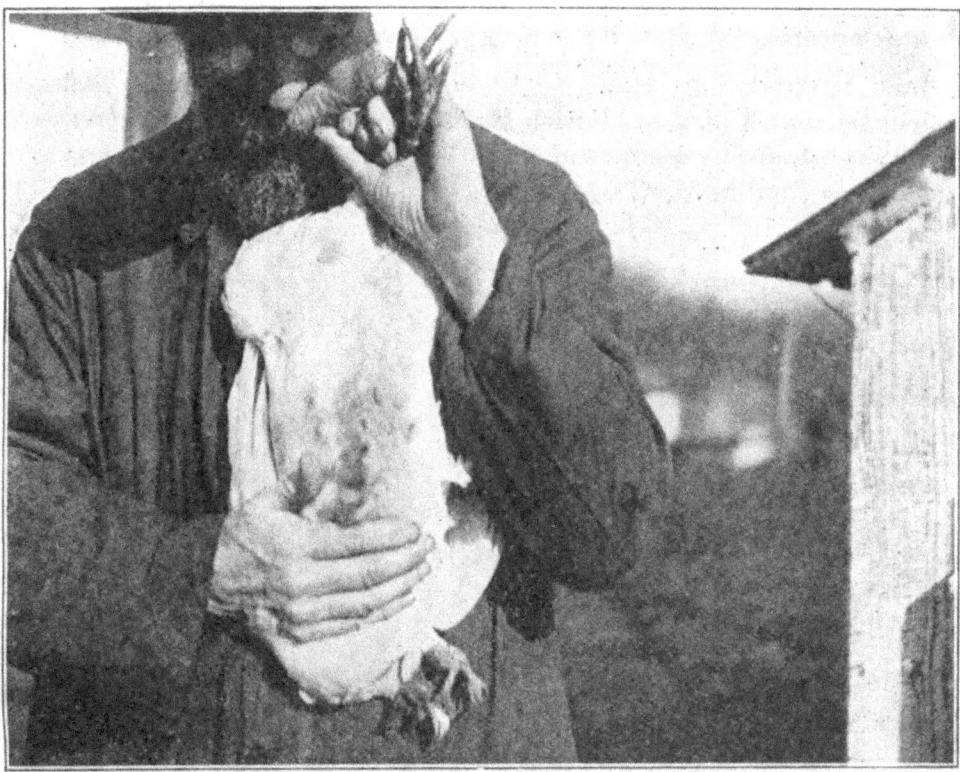

FIG. 9—Showing one movement that has proved an aid in testing type. The right hand is placed under the breast of the hen to steady her while the legs are drawn downward to bring the hen into position so that she may be examined for type (as in cut 10).

Fig. 9. After examining the hen as in Fig. 8, place the hand as in Fig. 9, and hold right hand firmly enough to prevent her from slipping down.

Fig. 10. Then move the left hand down, as in Fig. 10, and hold left hand firm enough to keep her in place while removing right hand.

Type.

Fig. 11. Now, brush feathers away from vent with back of hand and grasp end of pelvic bone so that it comes flush with outside of fingers, as in Fig. 11. This indicates the **Type** of the bird. Some will be one-sixteenth ($1/16$) of an inch thick, including the flank as held between the thumb and forefinger, as seen in Fig. 11, and will vary all the way up to one and a quarter ($1¼$) inches, including bone, gristle, fat, and flank, as seen in Fig. 31.

The reader is aware by this time that we are in the chapter pertaining to **Type,** the last of the three classes that it is necessary to divide poultry into in order to make a scientific classification to enable one to arrive at the approximate value of the "Individual Bird" as an **Egg** or **as a Meat** proposition (and without any regard as to its value as a breeder, which will be shown later). I wish to repeat here that **Type** is controlled wholly by temperament. We must select the temperament or combinations of temperaments that suit our purpose, and then, with the desired capacity and by scientific feeding, so

Fig. 10—Showing another movement that has proved an aid in testing type. The legs are drawn well under the hen, thus throwing the pelvic bones forward. The right hand is then removed and used to examine the thickness of the pelvic bones (Fig. 11).

as to keep the subject in proper **condition,** poultry-culture will become more of a science with the majority of poultrymen than it is at present. In order to prepare the reader for what is to follow, I will divide poultry into three distinct classes as to temperaments.

The hen that will produce the largest amount of **eggs** with the smallest amount of meat possible for her capacity is of the nervous temperament. The hen which uses one-half of her vitality in producing eggs and the other half of her vitality in producing meat—in other words, the dual-purpose hen—is a combination of both the sanguine and bilious temperaments and is called "the hen with the sanguine-bilious temperament." The hen that produces the largest amount of flesh and the smallest amount of eggs consistent with her capacity is of the lymphatic temperament.

Fig. 11—Shows method of testing type. The thumb and forefinger are placed one on each side of the pelvic bone so that you may estimate the thickness of the same, including flesh, fat, gristle, etc.

In a fowl all the different temperaments and their different degrees of combinations are indicated by the pelvic bones. In the horse they are indicated largely by the breed. The Arabian, the ideal running and trotting horse, is a good type of the nervous temperament, the coach horse is a good type of the sanguine-bilious temperament, and the Clyde is a good type of the lymphatic temperament. In cattle we have a good example of the nervous temperament in the Jersey, and of the lymphatic in the beef family of Durham, also Hereford and Polled Angus, while the Holstein and Ayrshire cattle are good types of the sanguine-bilious combined.

I have made this deviation so I could offer to my poultry friends this thought: that there are certain laws in Nature that have no regard for our theories, and the better we understand these laws, the less liable we are to make mistakes.

38   THE CALL OF THE HEN.

CHAPTER IV.

Capacity.

In the preceding chapters we have given the reader an idea of the method we use in judging the value of a hen for the pur-

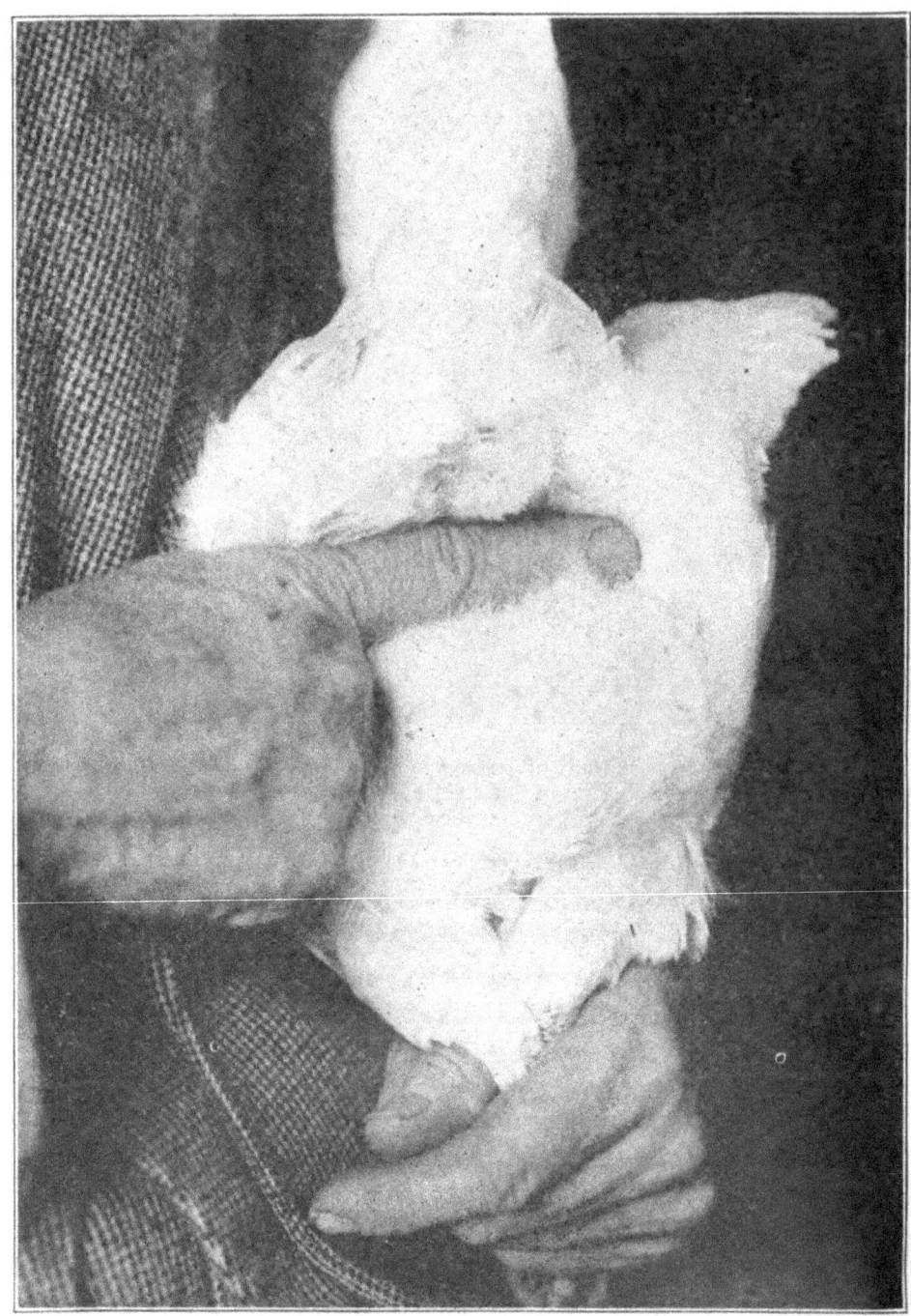

Fig. 12—One-finger abdomen. (Capacity.) This indicates a hen of very small capacity to consume and assimilate food. She never can be a large eater, hence of not much value.

pose we wish her for. In the succeeding chapters we will explain the method in detail.

First, we will take up "Capacity."

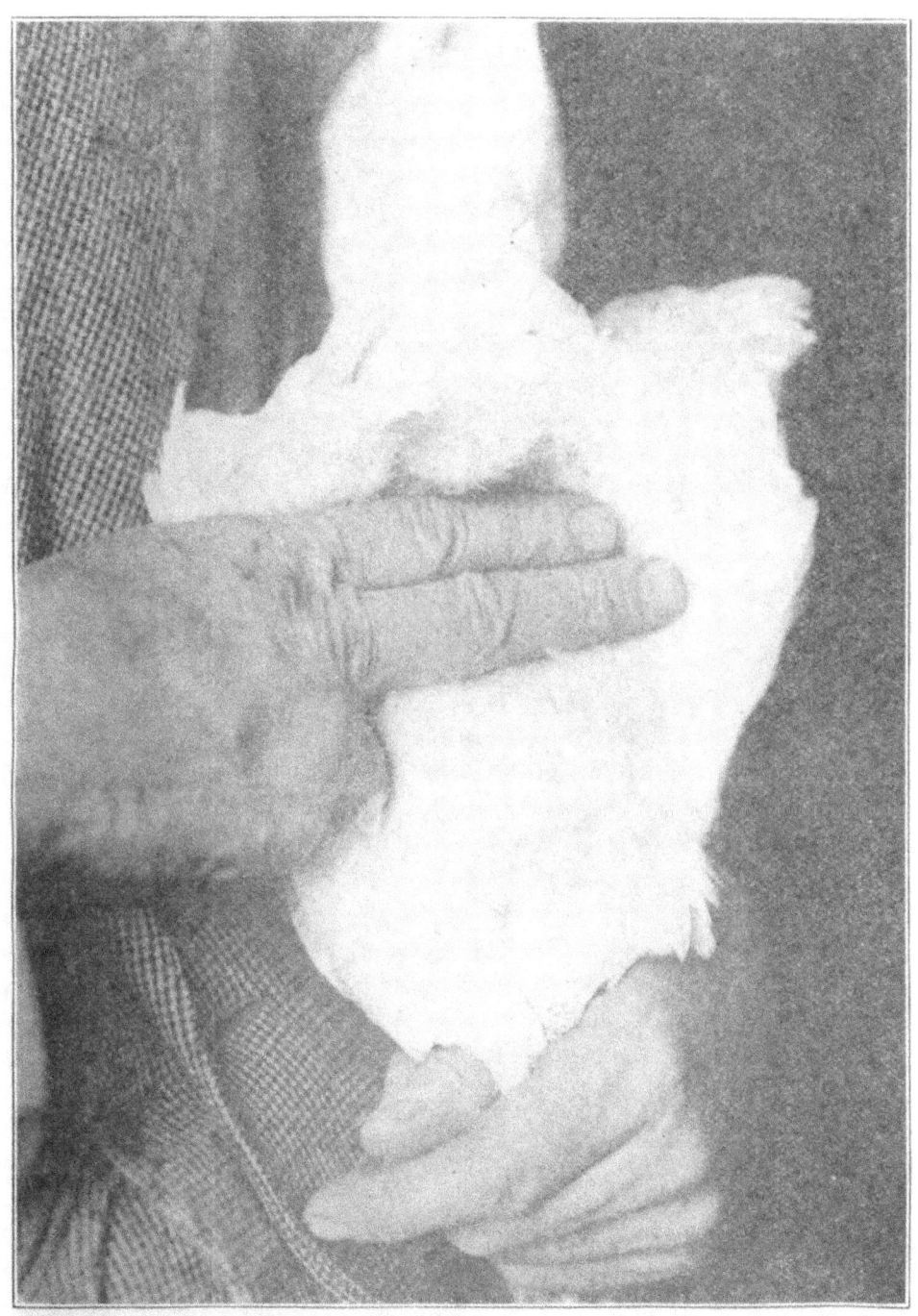

Fig. 13—Two-finger abdomen. (Capacity.) Slightly larger capacity than the preceding, but still of relatively small ability to consume food.

40   THE CALL OF THE HEN.

Fig. 12 shows a hen with only one finger capacity (¾ of an inch) between the two pelvic bones and the rear of the breast-bone.

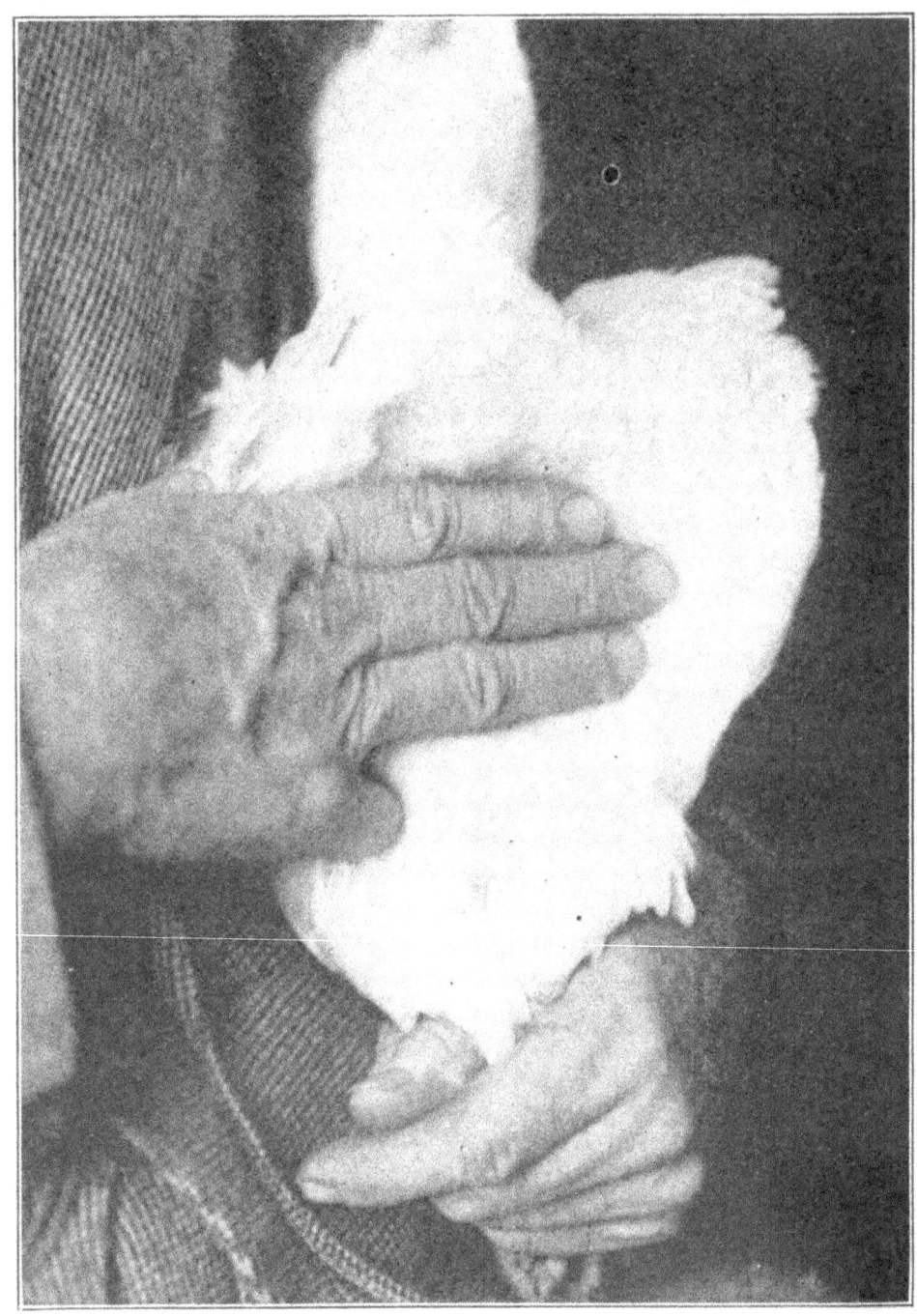

FIG. 14—Three-finger abdomen. (Capacity.) Indicating very good ability to consume and assimilate food. We find hens that lay as high as 180 eggs in their first laying year in this class, depending on the type.

Fig. 13 shows a hen with two fingers capacity (1½ inches) between the two pelvic bones and the rear of the breast-bone.

Fig. 14 shows a hen with three fingers capacity (2¼ inches) between the two pelvic bones and the rear of the breast-bone.

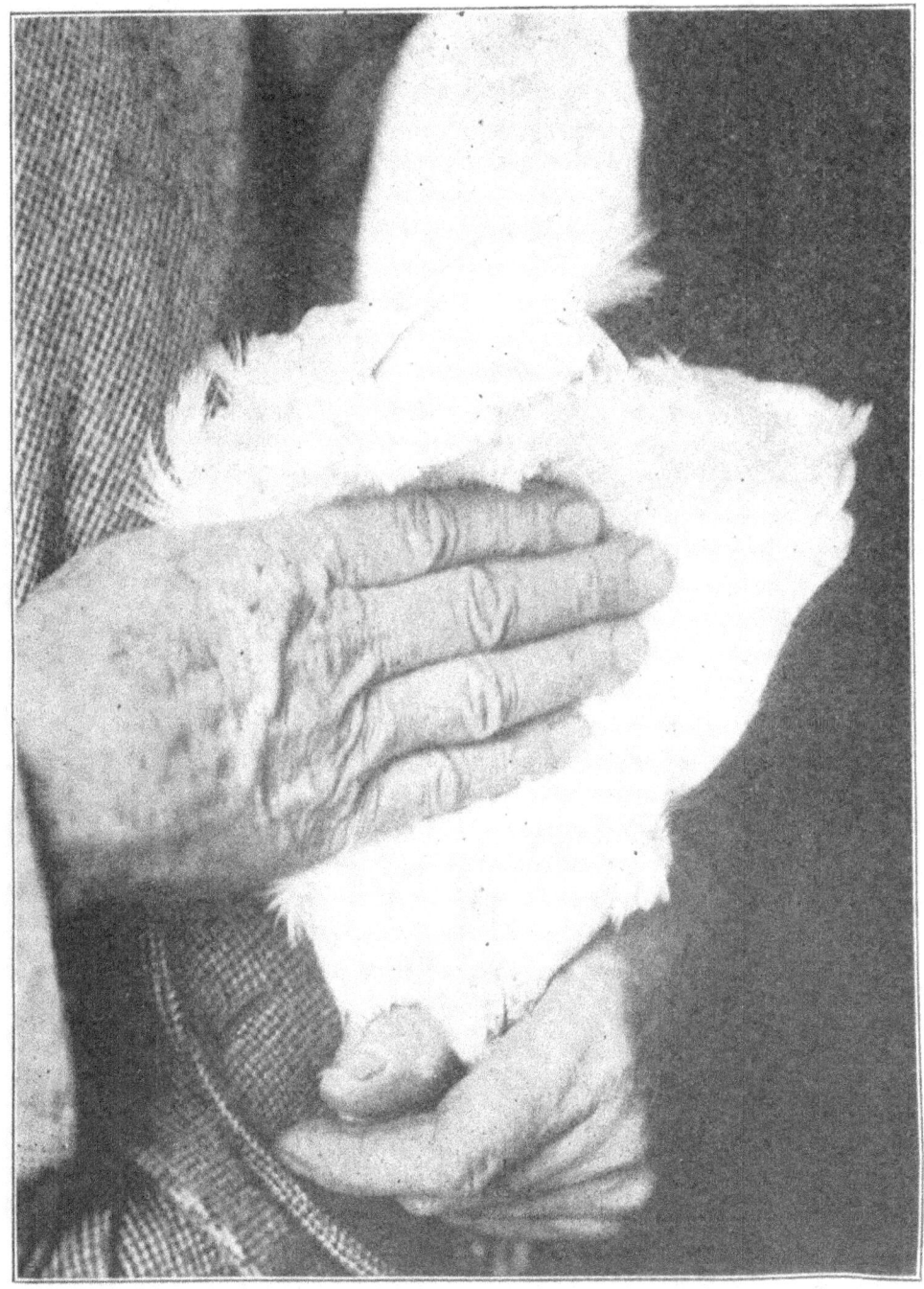

Fig. 15—Four-finger abdomen. (Capacity.) A hen of very large capacity to consume and assimilate food. We find 220-egg hens in this class, provided they have the right type.

Fig. 15 shows a hen with four fingers capacity (3 inches) between the two pelvic bones and the rear of the breast-bone.

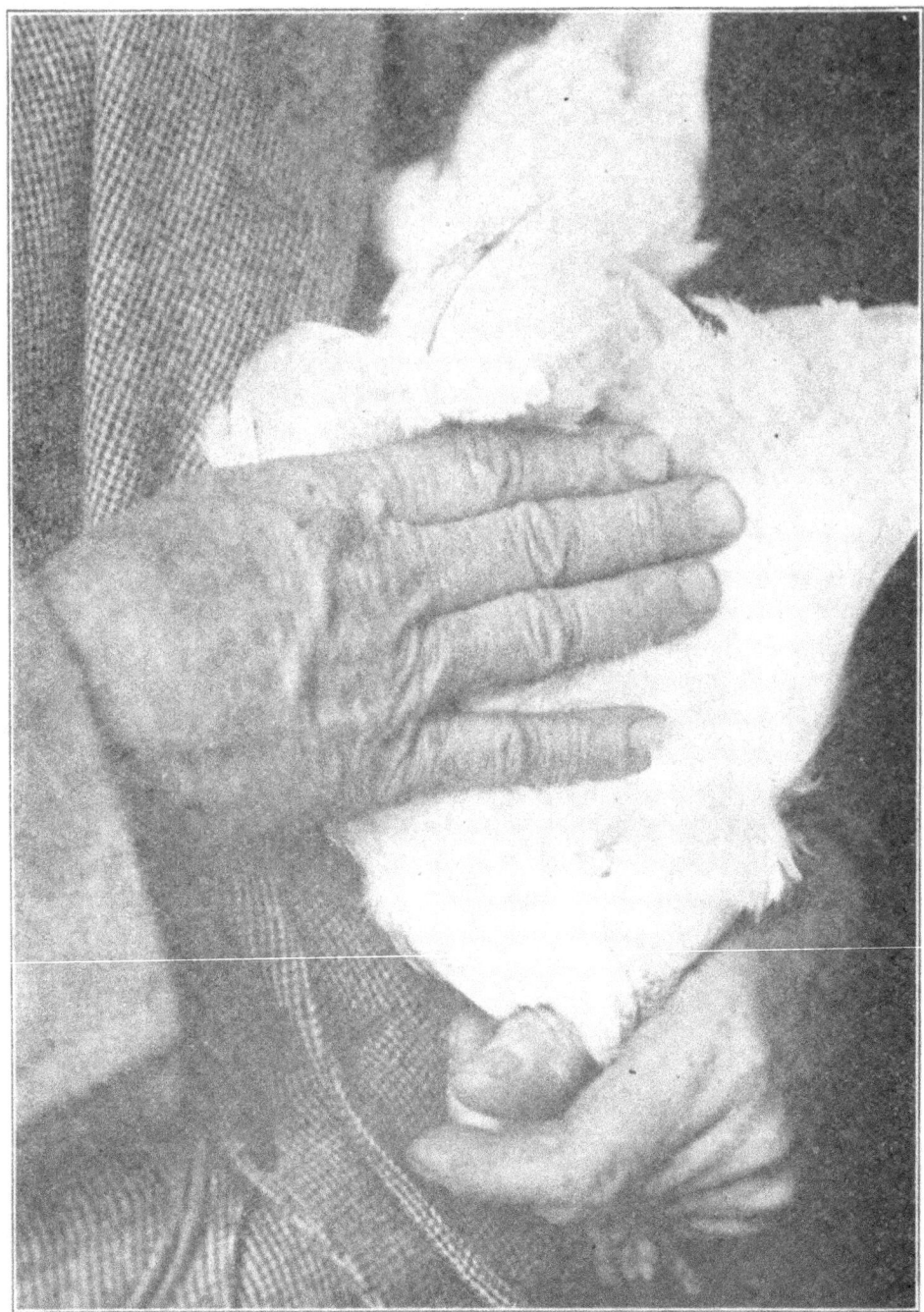

Fig. 16—Five-finger abdomen. (Capacity.) A hen of still larger ability to consume food than the preceding. We find 250-egg hens in this class, if of the right type.

Fig. 16 shows a hen with five fingers capacity (3¾ inches) between the two pelvic bones and the rear of the breast-bone.

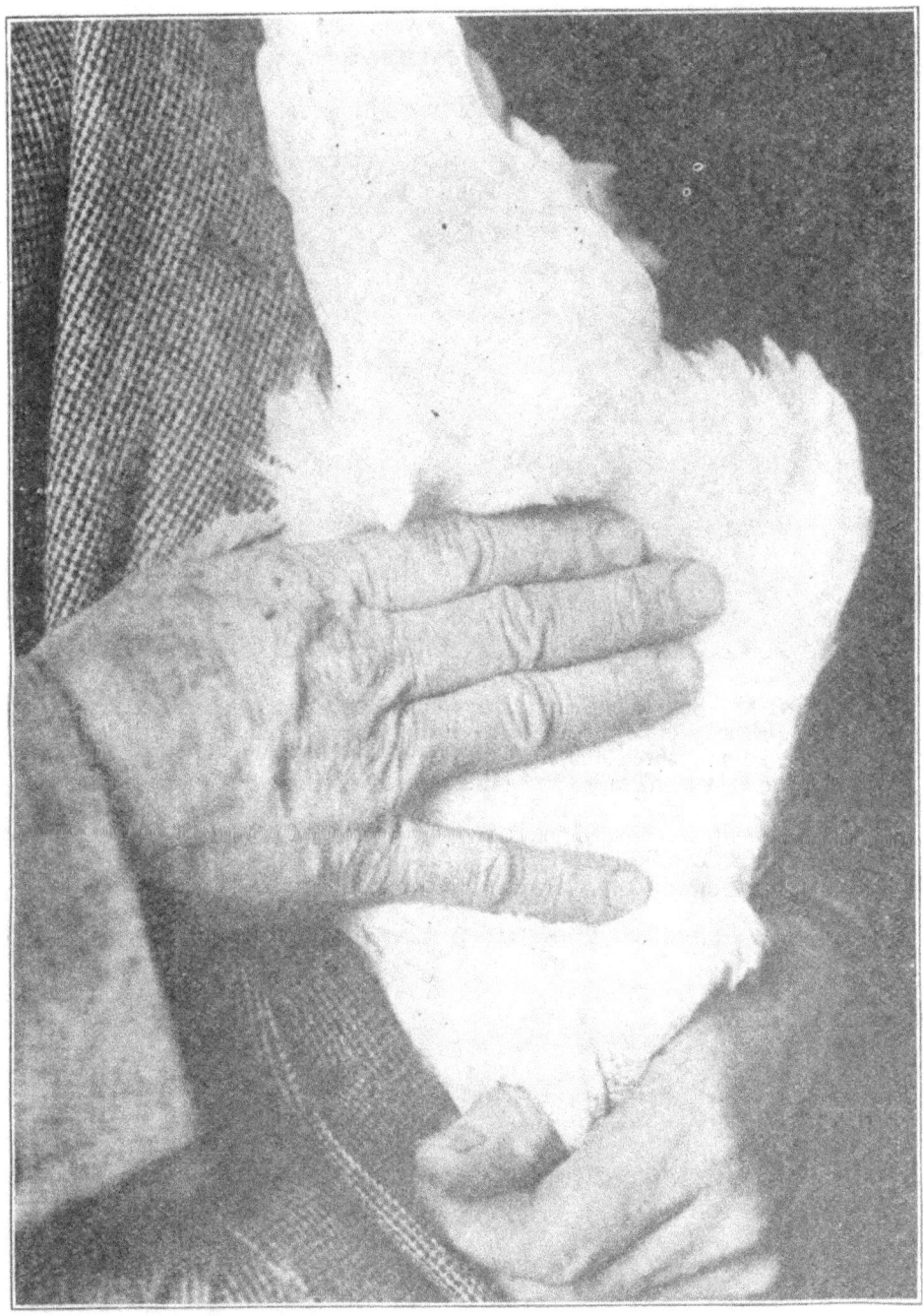

Fig. 17—Six-finger abdomen. (Capacity.) Indicating extremely large capacity to consume and assimilate food. She may be a 280-egg type hen or a heavy beef type hen.

Fig. 17 shows a hen with six fingers capacity (4½ inches) between the two pelvic bones and the rear of the breast-bone.

44 THE CALL OF THE HEN.

CHAPTER V.

Condition.

We next come to "Condition."

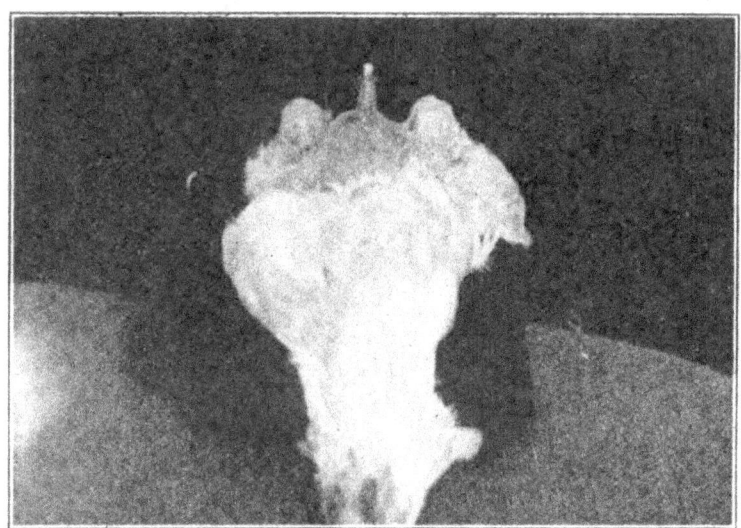

Fig. 18—Showing hen in very poor condition. The feathers being plucked away shows the actual condition of the flesh. We call a hen in this condition "three fingers out of condition," which indicates that her abdomen has shrunken up three fingers. If she now has a capacity of one finger, when in good condition she would be four fingers abdomen; if she has a capacity of two fingers now, she would have five fingers capacity when in good condition.

Fig. 18 shows a hen in very poor condition.

Fig. 19—Showing hen in good condition. You will note that the flesh is even with the breast-bone. This hen would show her normal abdominal depth when examined.

Fig. 19 shows a hen in perfect condition, as indicated by her full breast.

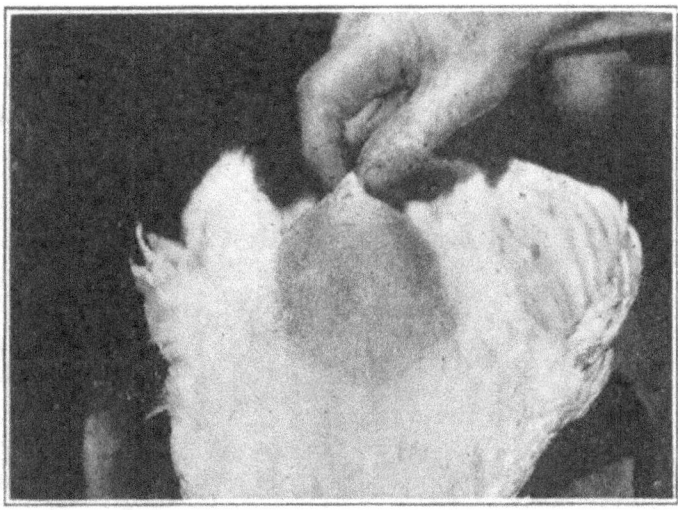

Fig. 20—Showing hen one finger out of condition. You will note that the flesh appears slightly shrunken away from the breast-bone. When the thumb and forefinger are placed as in the cut, about ½ inch from the front of the breast-bone, the flesh will be below the breast-bone, as shown by the mark on finger in Fig. 23. This would indicate that the hen was one finger less capacity. If three fingers now, she would be four fingers capacity when in condition, etc.

Fig. 20 is somewhat thinner, as indicated by breast-bone. We call her **one finger out of condition**.

The degrees of condition show the amount of shrinkage in abdominal depth. One finger out of condition shows she has shrunken one finger in depth of abdomen; two fingers out of condition shows she has shrunken two fingers in depth of abdomen; three fingers out of condition shows she has shrunken three fingers in depth of abdomen.

Fig. 21—Showing hen two fingers out of condition. The flesh is shrunken away from the breast-bone to about the depth indicated by the middle line on the finger in Fig. 23, which is about the middle of the first joint. This shows that she is two fingers less in abdominal depth than

46 THE CALL OF THE HEN.

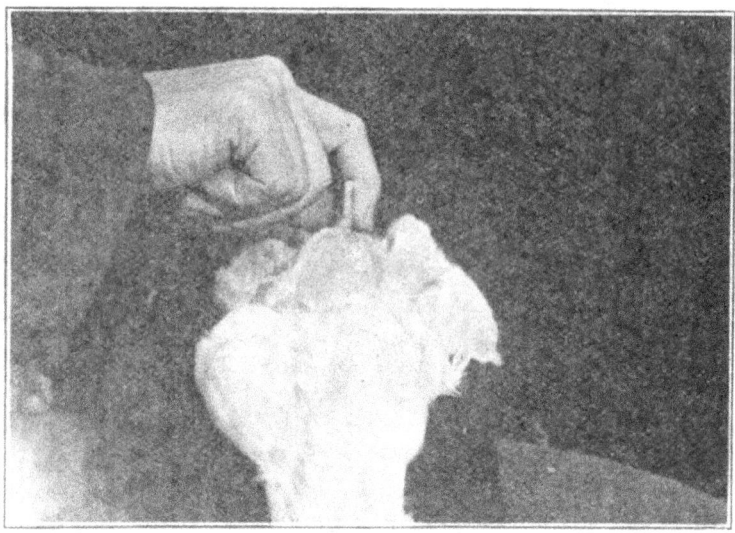

Fig. 22—Showing hen three fingers out of condition. This hen would be three fingers less in abdominal depth than when in good condition.

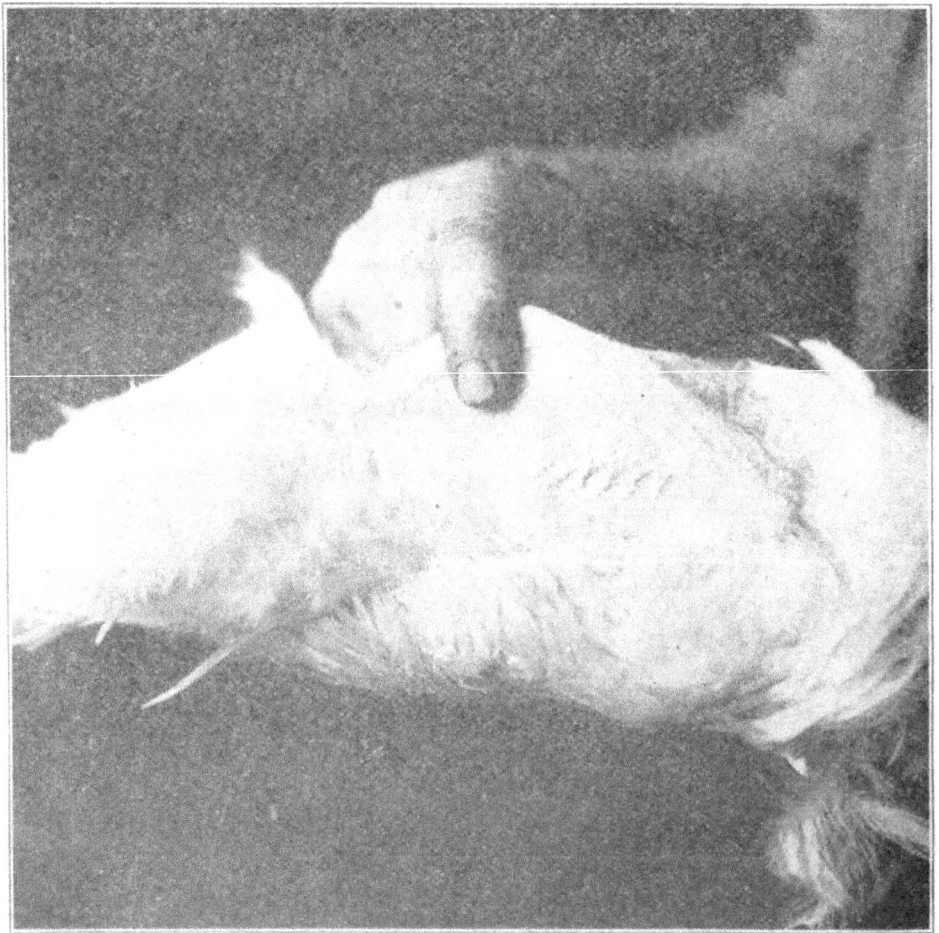

Fig. 22a—This shows you just where to place your finger on the keel

Fig. 21 is still thinner, as reader can see by the breast-bone. We call her **two fingers out of condition.**

Fig. 22 is still thinner. This we call **three fingers out of condition**, and is about as thin as a hen usually gets, if there is any chance for her ever being of any use.

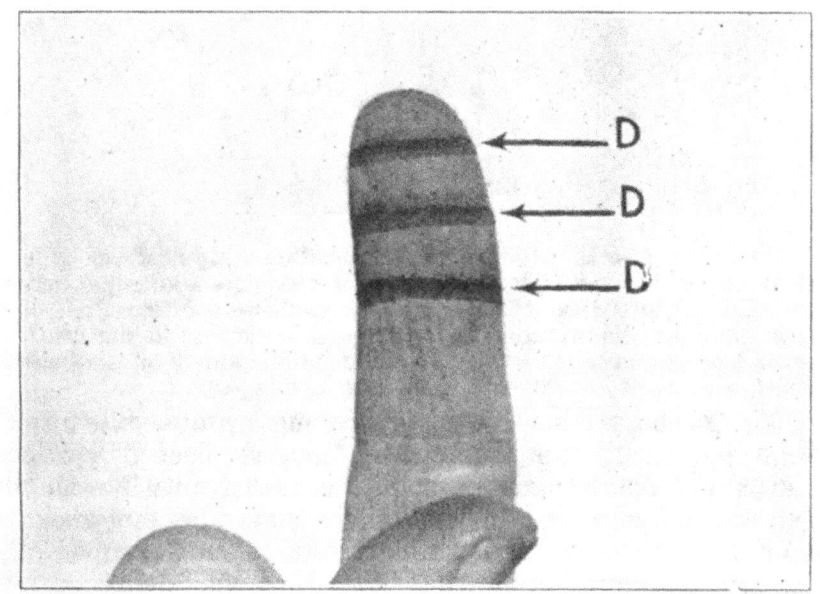

Fig. 23—Showing where the imaginary lines should be drawn on the first joint of the forefinger in order to judge the condition of the hen or pullet.

Fig. 23 shows about how the first joint of an index finger must be divided up to determine the three degrees of condition.

CHAPTER VI.

Type.

We now come to "Type." This is indicated by the thickness of the pelvic bones, together with the flesh, fat, gristle, and cartilage on same. (See page 17.)

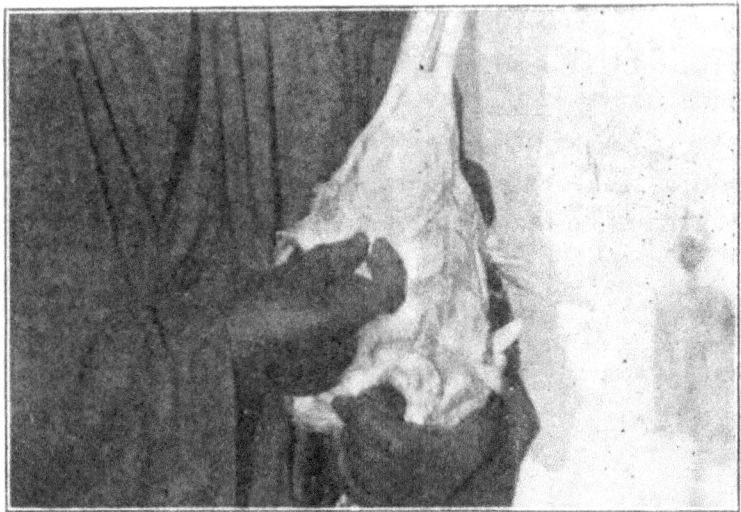

Fig. 24—1/16-inch pelvic bone. Indicating a typical egg-type hen, which means that virtually all the food she consumes above that necessary for bodily maintenance goes toward the production of eggs. If of one-finger abdomen, she would lay about 36 eggs in her first laying year; if of three-finger abdomen, she would lay about 180; and if of six-finger abdomen, she might lay 280 eggs in her first laying year.

Fig. 24 shows a hen whose pelvic bones are one-sixteenth ($1/16$) of an inch thick; that is about as thick as piece of cardboard that paper boxes are made of, and the reader must bear in mind that the measurement of the pelvic bone does not mean the bone alone, with the skin, flesh, gristle, and fat scraped off, as some may suppose, but with all the above included.

Fig. 25 shows a hen with pelvic bones one-eighth ($1/8$) of an inch thick.

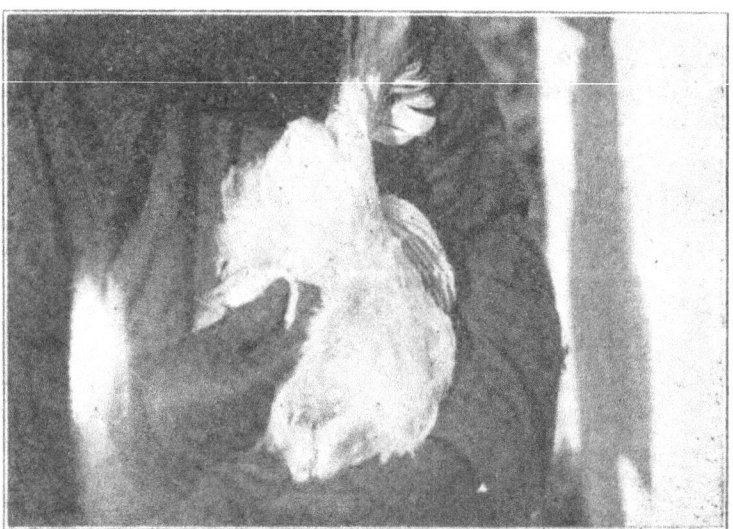

Fig. 25—1/8-inch pelvic bone; indicating egg type, but not so typical as the preceding. If of one-finger abdomen, she would lay about 32 eggs in her first laying year; if of three-finger abdomen, about 166 eggs; and if of six-finger abdomen, about 265 eggs in her first laying year.

FIG. 26—¼-inch pelvic bone; indicating a slightly more beefy hen than the preceding types, but still of the egg type. If of one-finger abdomen, she would lay about 24 eggs in her first laying-year; if of three-finger abdomen, about 138 eggs; and if of six-finger abdomen, about 235 eggs in her first laying year.

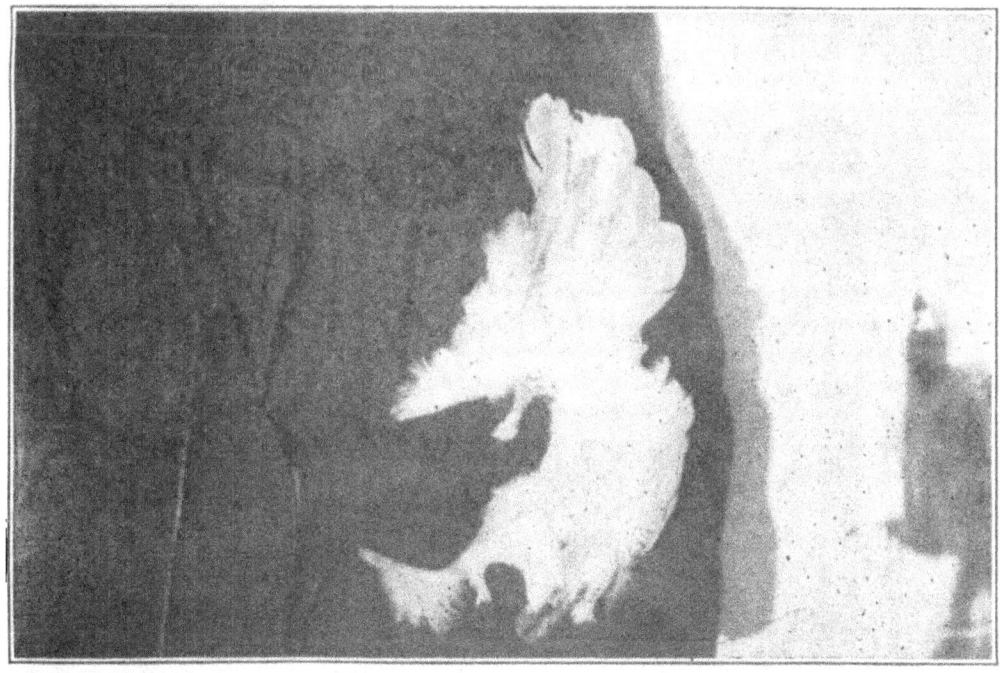

FIG. 27—⅜-inch pelvic bone; indicating that the hen uses a larger proportion of the food she consumes in making flesh and less in the production of eggs. A one-finger abdomen hen would lay about 16 eggs; a three-finger abdomen hen, about 110 eggs; and a six-finger abdomen hen, about 205 eggs in the first laying year.

Fig. 26 shows a hen with pelvic bones one-quarter (¼) of an inch thick.

Fig. 27 shows a hen with pelvic bones three-eighths (⅜) of an inch thick.

Fig. 28—½-inch pelvic bone; indicating a still more beefy hen than the preceding—that is, a still larger proportion of the food consumed is used to make flesh and less to produce eggs. If of one-finger abdomen, she would lay about 8 eggs; and if of three-finger abdomen, she would lay about 82 eggs; while if of six-finger abdomen, she would lay about 175 eggs in the first laying year.

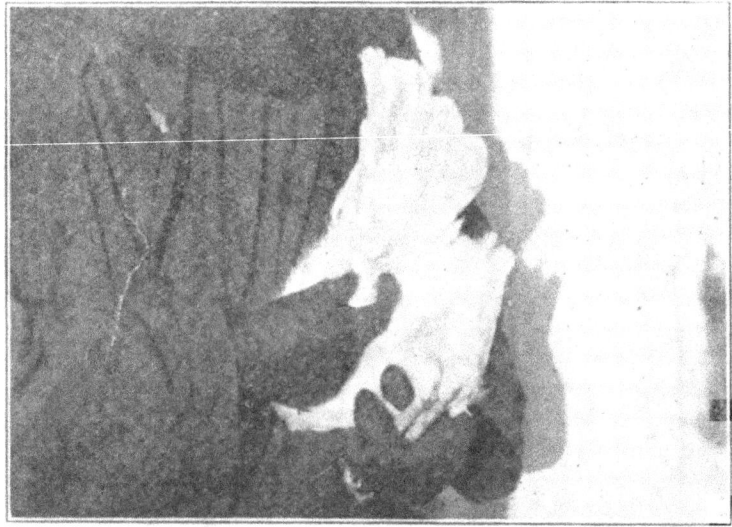

Fig. 29—¾-inch pelvic bone. A pretty good specimen of the beef type. We find no two-finger abdomen hens that have pelvic bones so thick, because they cannot consume enough food. A two-finger abdomen hen is virtually a non-layer; a three-finger abdomen hen will lay about 24 eggs; and a six-finger-abdomen hen will lay about 115 eggs in the first laying year.

Fig. 28 shows a hen with pelvic bones one-half (½) of an inch thick.

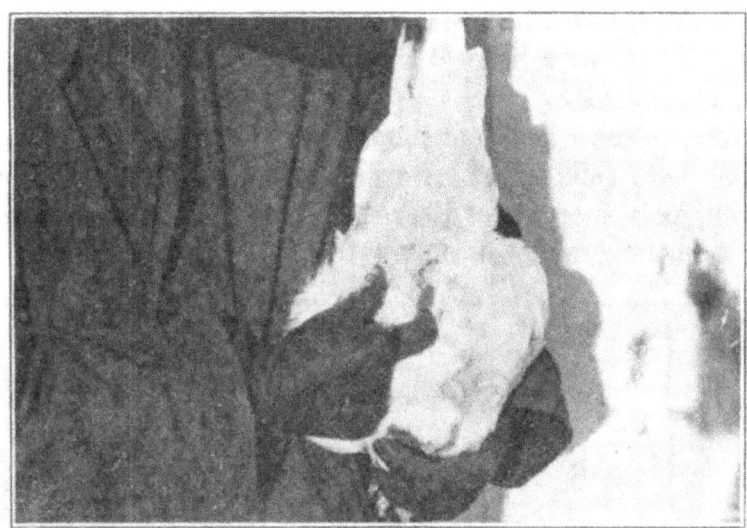

Fig. 30—1-inch pelvic bone. A very beefy type. Almost all the food consumed above that required for bodily maintenance is used in the making of flesh. We find them in the hens that have abdomens from four to six fingers deep. They lay very few eggs.

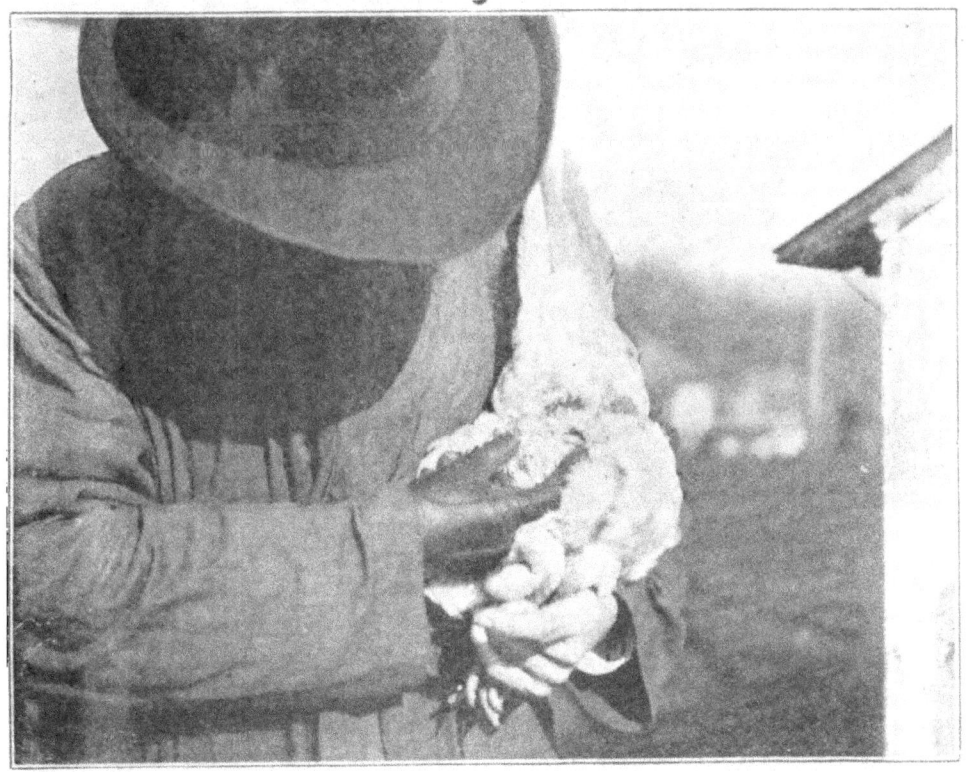

Fig. 31—1¼-inch pelvic bone. This indicates that the hen is of the typical beef type. She is an enormous feeder, hence only found in hens of about six-finger capacity. She will lay practically no eggs.

52 THE CALL OF THE HEN.

Fig. 29 shows a hen with pelvic bones three-quarters (¾) of an inch thick.

Fig. 30 shows a hen with pelvic bones one (1) inch thick.

Fig. 31 shows a hen with pelvic bones one and one-quarter (1¼) inches thick.

Now, please bear in mind that everything shown and related here refers to **Leghorns** and applies to other breeds as well, only in a lesser degree—so small that it amounts to almost nothing, as I will show later.

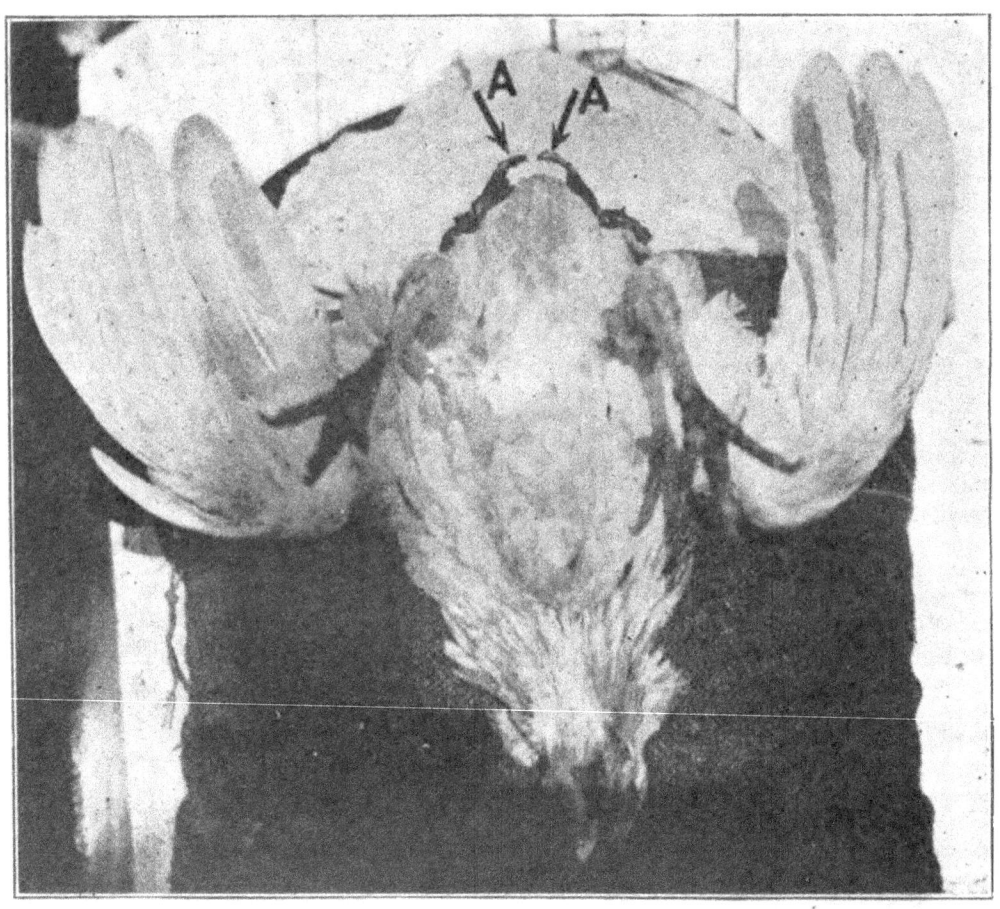

FIG. 32—Crooked pelvic bone. "A, A," Position No. 1.

A, A, Fig. 32, shows the pelvic bones with flesh cleaned off.

B, B, Fig. 33, shows the pelvic bones with flesh stripped off farther and painted black so they will show up better. You will notice that the pelvic bones in Fig. 32 and Fig. 33 are crooked. The majority of poultry have more or less crooked pelvic bones. Sometimes the bones come close together, which is an obstruction in laying, and should be bred away from as much as possible.

Fig. 34 shows perfect pelvic bones. In this form they

are very easy to take between the thumb and finger; also, when the hen wants to lay the vent has a chance to fall down between the pelvic bones, which allows the egg to be delivered without straining on the part of the hen. Not every poultry-

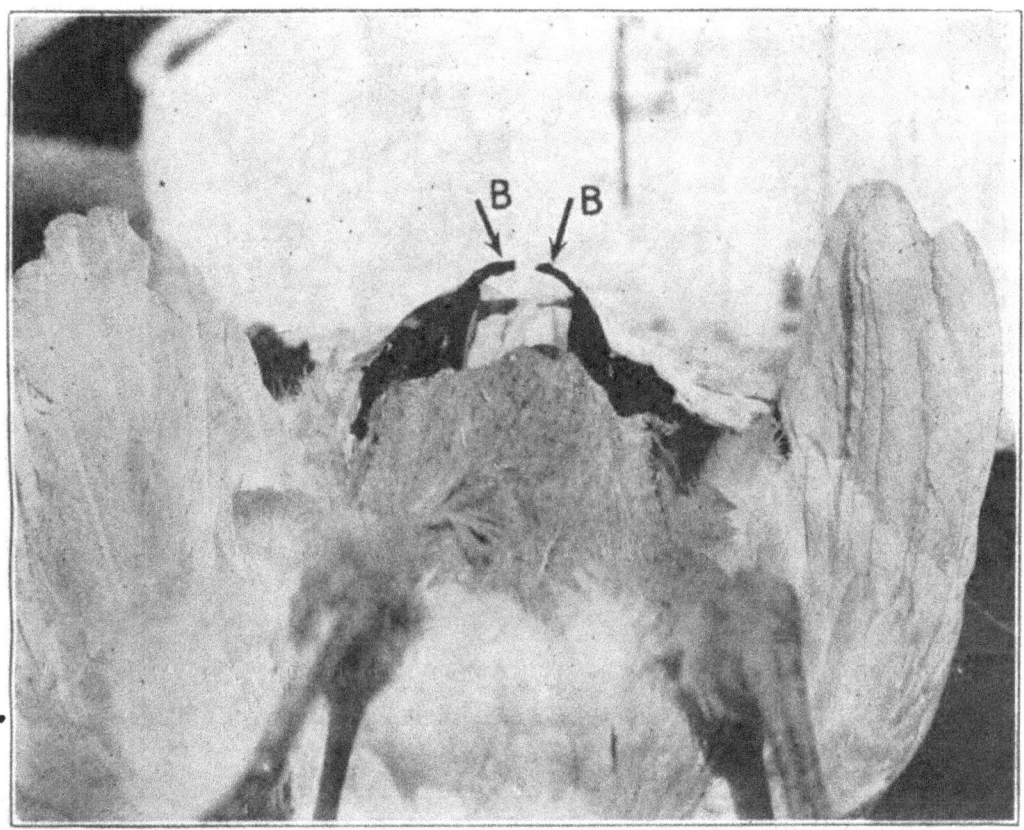

Fig. 33—Crooked pelvic bone, "B, B," Position No. 2. Hens with bones curved like this will lay about 20 per cent less than hens of the same type and capacity with straight pelvic bones, as in Fig. 34.

man, but every poultrywoman has seen cases where a hen has gone on the nest and after a couple of hours commenced to cackle her head off. Presently we hear the whole flock take up the chorus, and going to see what the trouble is, we find the hens holding an "Old Maids' Convention" and declaring they will never lay another egg, it hurts them so much to do so. On examining them, we find the pelvic bones so crooked they come together like the horns on a Jersey cow, and when the hens lay, instead of the vent dropping down between the pelvic bones, allowing the egg to be released in an easy manner in a few minutes after the hen goes on the nest, the egg is forced to be delivered between the pelvic bones and tail bone, thus prolonging the agony of the hen sometimes for hours, when,

54   THE CALL OF THE HEN.

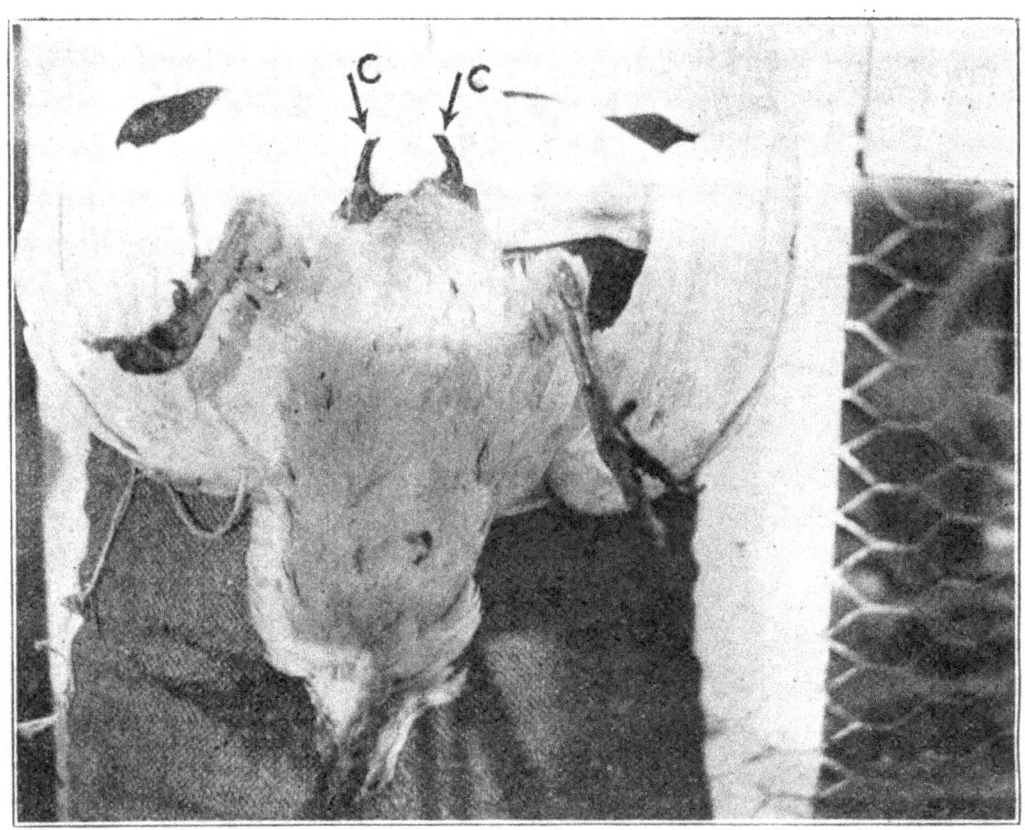

Fig. 34—Most perfect pelvic bones, "C, C." Hens with pelvic bones like this will lay about 20 per cent more than those having bones like Fig. 33.

if she was built right, as in Fig. 34, she would be relieved of the egg without pain in a few minutes. And instead of wasting vitality in getting relieved of the egg, she would be rustling around for material to build another one, and thus add at least 20 per cent to her egg-producing value. This matter of crooked pelvic bones is more frequent in some breeds than in others, and is a serious matter that is very easily remedied by breeding only from birds with the straightest pelvic bones; especially looking after the male birds, as one male bird with crooked pelvic bones will transmit this defect to all of his daughters.

When I came to Petaluma, I found whole flocks of thousands of hens with crooked bones; now they are very rare. The poultry-breeders soon caught on to my straight-and-thin-pelvic-bone idea; and I think the Society for the Prevention of Cruelty to Animals should recognize my services in relieving millions of hens of the agony of parturition.

The reader will please bear in mind that Fig. 34 represents 100 per cent pelvic bone and holds the same relation to pelvic bones in general that a bird that scores 100 in the show-room holds to all other high-class birds.

A 250-egg type cock bird or cockerel with pelvic bones like Fig. 34 would be of inestimable value. The writer has cock birds like the above that he would not part with for any money, for the reason that it would take several years' breeding to produce their equals.

If the reader has male birds whose pelvic bones are far enough apart that he can grasp the ends with thumb and finger when measuring the thickness, he should be satisfied until he can do better.

I have found more straight pelvic bones in some strains of Orpingtons than in any other breed. So long as the pelvic bones are comparatively straight after leaving the frame and do not curve abruptly toward the ends, the birds may be used as breeders, with the assurance that some of the offspring will show a wonderful improvement in this respect. Figs 32, 33, and 34 are extreme cases.

CHAPTER VII.

The First Laying Year.

What is meant by "the first laying year"? All old poultrymen know what the above means, and I have no doubt some of my readers may be impatient with me for explaining little things that are so familiar to them; but they will remember that poultry parlance is not all contained in the dictionary, and a great deal of the contents of this book may be Greek to the beginners in the poultry business who will read this work. For this reason I cannot be too plain in my language or too careful of details in explaining matters. The first laying year has nothing whatever to do with the age of a hen or pullet. I have had hens that had passed their first laying-year when they were sixteen months old. On the other hand, I have seen hens that were over four years old that had not commenced on their first laying year. The hen that had passed her first laying year when she was sixteen months old had commenced to lay when she was four months old, while the hens that were over four years old had never laid an egg. So the reader will see the first laying year commences with the first egg a pullet lays and ends one year from that date, when her second laying year commences. Some pullets will commence to lay at four months old, while others of exactly the same type, fed and cared for in the same manner, will not lay before they are eight months old, owing to different environment. Everything else being equal, poultry will develop faster on a warm, dry, sandy soil than they will on a black, damp, heavy soil; and they will mature much sooner in a good corn country, where it is warm in the shade and warm at night, than they will in a poor corn country, where it is cool at night and cool in the daytime in the shade. I have raised Leghorn pullets that were fully developed in size and form and laid a full-sized egg when they were four months old.

It can be done in Massachusetts, New York, New Hampshire, and Minnesota, and in parts of California, where the nights are so warm that one can sleep comfortably under a sheet only; but not where you have to cuddle under a lot of blankets on a summer night to keep warm.

CHAPTER VIII.

The Selection of Types.

If the reader has practiced handling a hen as in Figs. 5, 6, 7, 8, 9, 10, and 11, we will proceed with a lesson in judging hens as to the number of eggs they will lay their first laying year.

We will look for a small hen to commence with, as she will be easier to handle. Having our hen, we will hold her as

CHART 1.

One-finger Abdomen.

$1/16$ pelvic bone................36 eggs
$1/8$ pelvic bone................32 eggs
$3/16$ pelvic bone................28 eggs
$1/4$ pelvic bone................24 eggs
$5/16$ pelvic bone................20 eggs
$3/8$ pelvic bone................16 eggs
$7/16$ pelvic bone................12 eggs
$1/2$ pelvic bone................ 8 eggs
$9/16$ pelvic bone................ 4 eggs
$5/8$ pelvic bone................ 0 eggs

nearly as we can as in Fig. 5, and try to have her head as in Fig. 6, so she can see nothing. She will then be easier to handle. Place hand across her abdomen, as in Fig. 7. She may be a one-finger-abdomen hen, as in Fig. 12. Then hold her as in Fig. 8. Her breast may be as in Fig. 19; if so, she will be in good condition. Next go through movements as in Figs. 9 and 10 and hold her and examine her pelvic bone as in Fig. 11. Her pelvic bone may be one-sixteenth ($1/16$) of an inch thick, as in Fig. 24. Now look on Chart 1. Your hen is one finger abdomen, in good condition, and her pelvic bone is one-sixteenth ($1/16$) of an inch thick. You will see that she is a 36-egg type hen. That means that if this hen is one of a large number on a commercial poultry plant, she is capable of laying three dozen eggs her **first laying year,** if she is fed and cared for properly, barring accidents and disease. So we will call her a 36-egg type hen.

We will drop this hen and take another from the crate and go through the same movements. Hold her as in Fig. 5 or Fig. 7, with head as in Fig. 6 (she may also be a one-finger-abdomen hen, as in Fig. 12), then examine for condition, as in Fig. 8. Her condition may be good, as in Fig. 19; then hold as in Figs. 9 and 10, and measure thickness of pelvic bone, as in Fig. 11. Her pelvic bone may be three-eighths ($3/8$) of an inch thick, as in Fig. 27; in that case she would read like this: One-finger abdomen; good condition; three-eighths ($3/8$) pelvic bone. Now, look on Chart 1, and you will find she is a 16-egg type hen.

We will drop her and take another from the crate, and go through the same movements as before. This hen may be a one-finger-abdomen hen also, in good condition, with pelvic bone $1/2$ inch thick, as in Fig. 28, and by consulting Chart 1 we find she is an 8-egg type hen.

We drop her and take another from the crate. She may be a hen with one-finger abdomen, as in Fig. 12. When we examine her condition, we find she is like Fig. 20, which indicates that she is one finger out of condition (the subject of "Condition" is explained in Chapter V.); her pelvic bone may be $1/16$ of an inch thick, as in Fig. 24. This hen will read different from the other hen that was $1/16$ pelvic bone. This hen is out of condition. She may have been in condition up to a few weeks previous to our examination of her; the cause of her lack of condition may be improper food or care, or both, or it may be due to moulting, or she may have been broody. In any of these cases it would not be the hen's fault that she was out of condition, and she should not be held responsible for it. Her condition indicates that there is something wrong, and it's up to her owner to right the wrong, and when we do right the wrong, the hen will come back into condition, and her abdomen will then measure two fingers instead of one finger. We must, therefore, read her as a two-finger-abdomen hen, $1/16$-inch pelvic bone, when, by looking on Chart 2, we find her capacity would be 96 eggs her first laying year, if we kept her in condition.

We will drop her, and take another hen out of the crate. This hen may be a one-finger-abdomen hen, as in Fig. 12. When we examine her for condition, we find her as in Fig. 21; this indicates that she is two fingers out of condition; her pelvic bone may be $1/16$ of an inch. Under her present condition, she might lay 36 eggs her first laying year, whereas, if she were kept in good condition, she might lay 180 eggs.

We will drop this hen and take up another one. She may be two fingers abdomen and her breast-bone may be as in Fig. 19. Her pelvic bone may be $1/16$ of an inch. We would read

CHART 2.

Two-finger Abdomen.

$1/16$ pelvic bone................96 eggs
$1/8$ pelvic bone................87 eggs
$3/16$ pelvic bone................78 eggs
$1/4$ pelvic bone................69 eggs
$5/16$ pelvic bone................60 eggs
$3/8$ pelvic bone................51 eggs
$7/16$ pelvic bone................42 eggs
$1/2$ pelvic bone................33 eggs
$9/16$ pelvic bone................24 eggs
$5/8$ pelvic bone................15 eggs
$11/16$ pelvic bone................ 6 eggs
$3/4$ pelvic bone................ 0 eggs

her as a two-finger-abdomen hen in good condition, pelvic bones $1/16$ of an inch thick. We will look on Chart 2 at $1/16$-inch pelvic bone, and find she is a 96-egg type hen.

We will drop her and take another from the crate. She may be two fingers abdomen and two fingers out of condition, as in Fig. 21, with pelvic bones ¼ of an inch thick. She would read two fingers abdomen and two fingers out of condition. She would be four fingers abdomen if in condition, and ¼-inch pelvic bones. Being a four-finger-abdomen hen (if in condition), we will look on Chart 4 at ¼-inch pelvic bone, and find she is a 175-egg type hen. We will drop her.

Take another. She may be a two-finger-abdomen hen, as in Fig. 13, in good condition, as in Fig. 19, with pelvic bones ¾ of an inch thick, as in Fig. 29. She would read two fingers abdomen; good condition; ¾-inch pelvic bones. We will look on Chart 2 for ¾-inch pelvic bone, and find this hen will lay nothing. This does not mean that she is an absolutely barren hen, that she will never lay an egg (I will explain this when we get to the six-finger-abdomen hen); she may lay a few, perhaps half a dozen, in the spring when the crows lay; but as a **commercial** proposition she will have no more value than the hen that never laid an egg. Everything she consumes goes to the making of flesh, except what she uses in bodily maintenance.

We will drop her and take another. She may be a three-finger-abdomen hen, as in Fig. 14. Her condition may be as in Fig. 19, with pelvic bones as in Fig. 24. She would read

CHART 3.

Three-finger Abdomen.

$1/16$ pelvic bone..............180 eggs
$1/8$ pelvic bone..............166 eggs
$3/16$ pelvic bone..............152 eggs
$1/4$ pelvic bone..............138 eggs
$5/16$ pelvic bone..............124 eggs
$3/8$ pelvic bone..............110 eggs
$7/16$ pelvic bone.............. 96 eggs
$1/2$ pelvic bone.............. 82 eggs
$9/16$ pelvic bone.............. 68 eggs
$5/8$ pelvic bone.............. 54 eggs
$11/16$ pelvic bone.............. 40 eggs
$3/4$ pelvic bone.............. 26 eggs
$13/16$ pelvic bone.............. 12 eggs
$7/8$ pelvic bone.............. 0 eggs

three fingers abdomen; in good condition; $1/16$-inch pelvic bone. We look on Chart 3 at $1/16$-inch pelvic bone, and find that this hen is a 180-egg type.

We will drop her and take another. She may be another three-finger-abdomen hen, like Fig. 14; she may be in good condition, like Fig. 19, and her pelvic bone may be ½ inch thick, like Fig. 28. She would read three fingers abdomen; good condition; ½-inch pelvic bone. We will look on Chart 3 at ½-inch pelvic bone, and find this hen is an 82-egg type hen.

We will take another hen. She may be three fingers abdomen, like Fig. 14; she may be in good condition, like Fig. 19, and her pelvic bone may be ¾ of an inch thick, as in Fig. 29. We will read her as a three-finger-abdomen hen; in good condition; ¾-inch pelvic bone. We will look on Chart 3 at ¾-inch pelvic bone, and find she is a 26-egg type hen.

We will pick up another hen. She may be three fingers capacity, as in Fig. 14; she may be three fingers out of condition, as in Fig. 22, and her pelvic bones may be $1/16$ of an inch thick, as in Fig. 24. We would read this hen as three fingers abdomen; three fingers out of condition; and $1/16$-inch pelvic bone. When a hen is three fingers out of condition she is in a serious way. She may have been setting or laying heavily and have been underfed. In either case, good care and plenty of the right kind of feed will bring her back into condition, provided she has not contracted tuberculosis (going light) or some other wasting disease. I will cite a couple of cases out of hundreds that have come under my observation.

One was a Barred Rock hen that I intended to set on duck eggs; she was six fingers abdomen, in good condition when I put her on the nest, and ¼-inch pelvic bones; that indicated that she was a 235-egg type hen. She was on the nest two weeks before the duck eggs arrived and four weeks on the duck eggs, making six weeks setting. Owing to stress of other work, and being confined in an out-of-the-way place, she was somewhat neglected, and when the ducklings were hatched she was three fingers abdomen and three fingers out of condition, thus indicating a 138-egg type hen. Six weeks later she was laying, and had developed to six fingers abdomen, which was her normal condition.

Another case was where a gentleman was in a class that took instructions. After the close of the meeting he brought a hen that was three fingers out of condition. He said that was his best hen, and asked me how many eggs she would lay. She was three fingers abdomen, three fingers out of condition, and $1/16$-inch pelvic bone. Her head and actions indicated perfect health. I told him she might lay 180 eggs her first laying year, if her condition had been the same as it was at the present 'time; but if she was my hen I thought I might be able to make her lay 280 eggs. "You don't feed her half enough." He replied, "That is the only hen I have that lays a white egg. I got her when a pullet, before she commenced to lay. She has been laying about a year and has laid 176 eggs. I had a small lot of hens at the time that were so fat they were dying, and I cut down their feed and have fed them sparingly ever since, so they would not get too fat and die." I went to his place, and found he had three types of hens: the typical meat type (one with pelvic bones $1\frac{1}{8}$ inches thick), some with pelvic bones $\frac{1}{2}$ inch thick, and this hen that laid the white eggs, whose pelvic bones were $\frac{1}{16}$ of an inch thick. I told him to segregate his hens into three lots, and feed them according to their type. Give the egg-type hens all the grain they could clean up each day in the scratching-shed, with a dry balanced mash before them all the time; the dual-purpose hens should be fed all the grain they wished to scratch for, with an occasional mash; and the beef-type hens should be fed what grain they could clean up in the scratching-shed in about an hour. The litter should be good and deep in all cases. I did not mention charcoal, grit, shells, and green stuff, as that is not my business. Every man who takes a poultry paper knows that part of the business, and every person who keeps poultry should take a poultry paper in order to keep posted on current poultry topics.

The gentleman wrote me over a year later that he had succeeded in bringing the hen up to normal condition, as in Fig. 19, but after laying awhile she went back to five fingers abdomen and one finger out of condition, and had laid 238 eggs her next laying year.

We will now take another hen. She may be four fingers abdomen, as in Fig. 15, in good condition, as in Fig. 19, and her pelvic bones may be $\frac{1}{16}$ of an inch thick, as in Fig. 24. She would read four fingers abdomen; good condition; $\frac{1}{16}$-inch pelvic bone. If we consult Chart 4, we will find she is a 220-egg type hen.

The next hen may be also four fingers abdomen, as in Fig. 15, in good condition, as in Fig. 19, with pelvic bones $\frac{1}{2}$ inch, as in Fig. 28. She would read four fingers abdomen; in good condition; $\frac{1}{2}$-inch pelvic bones. We will see by Chart 4 that this is a 115-egg type hen.

Our next hen may be a four-finger abdomen hen; condition good; pelvic bones 1 inch thick. We would read her as four fingers abdomen; condition good; pelvic bones 1 inch. If we look on Chart 4 at 1-inch pelvic bones, we will find this hen will lay approximately nothing.

CHART 4.
Four-finger Abdomen.

$1/16$ pelvic bone............220 eggs
$1/8$ pelvic bone............205 eggs
$3/16$ pelvic bone............190 eggs
$1/4$ pelvic bone............175 eggs
$5/16$ pelvic bone............160 eggs
$3/8$ pelvic bone............145 eggs
$7/16$ pelvic bone............130 eggs
$1/2$ pelvic bone............115 eggs
$9/16$ pelvic bone............100 eggs
$5/8$ pelvic bone............ 85 eggs
$11/16$ pelvic bone............ 70 eggs
$3/4$ pelvic bone............ 55 eggs
$13/16$ pelvic bone............ 40 eggs
$7/8$ pelvic bone............ 25 eggs
$15/16$ pelvic bone............ 40 eggs
1-in. pelvic bone............ 0 eggs

Our next hen may be a four-finger-abdomen hen, one finger out of condition, $1/8$-inch pelvic bone. She would indicate a 205-egg type hen under her present condition, but we would read her four fingers abdomen, one finger out of condition; that would mean a five-finger-abdomen hen if in condition, $1/8$-inch pelvic bone. We look on Chart 5 at $1/8$ pelvic bone, and find she is a 235-egg type hen.

Our next hen may be a five-finger-abdomen hen, as in Fig. 16; she may be in good condition, as in Fig. 19, and her pelvic bones may be $1/16$ of an inch, as in Fig. 24. She will read five fingers abdomen; condition good; pelvic bones $1/16$ inch. We look on Chart 5 at $1/16$ pelvic bone, and find she is a 250-egg type hen.

Our next hen may be a five-finger-abdomen hen, as in Fig. 16; she may be in good condition, as in Fig. 19, and her pelvic bones may be $3/8$ inch thick, as in Fig. 27. We would read her as five fingers abdomen; good condition, and $3/8$-inch pelvic bones. Chart 5 would show us that she was a 175-egg type hen.

CHART 5.

Five-finger Abdomen.

$1/16$ pelvic bone..............250 eggs
$1/8$ pelvic bone..............235 eggs
$3/16$ pelvic bone..............220 eggs
$1/4$ pelvic bone..............205 eggs
$5/16$ pelvic bone..............190 eggs
$3/8$ pelvic bone..............175 eggs
$7/16$ pelvic bone..............160 eggs
$1/2$ pelvic bone..............145 eggs
$9/16$ pelvic bone..............130 eggs
$5/8$ pelvic bone..............115 eggs
$11/16$ pelvic bone..............100 eggs
$3/4$ pelvic bone.............. 85 eggs
$13/16$ pelvic bone.............. 70 eggs
$7/8$ pelvic bone.............. 55 eggs
$15/16$ pelvic bone.............. 40 eggs
1-in. pelvic bone.............. 25 eggs
$1 1/16$ pelvic bone.............. 10 eggs
$1 1/8$ pelvic bone.............. 0 eggs

The next hen may be a five-finger-abdomen hen; condition good; pelvic bones 1 inch thick. She would read five fingers abdomen; good condition; 1-inch pelvic bones. The chart would indicate that she was a 25-egg type hen.

The next hen may be a six-finger-abdomen hen, as in Fig. 17; she may be in good condition, and her pelvic bones may be $1 1/4$ inches thick as in Fig. 31. I hear the reader say, "What breed of a hen has pelvic bones as thick as that? or do you mean that both of her pelvic bones are $1 1/4$ inches thick, counting them both together?" No; I mean that each one of her pelvic bones is $1 1/4$ inches thick. Counting the bone, gristle, fat, and flesh (flank), both of the pelvic bones would be $2 1/2$ inches thick. When we speak of pelvic bones being so and so thick, we always mean one of them. And as to breed, this hen is a Single Comb White Leghorn; she is the typical beef type. You will see by Chart 6 that she will lay practically nothing; and here I will explain this matter.

A man once brought me a two-and-a-half-year-old hen that he had trap-nested for two years, and asked me to tell him how many eggs she had laid her first laying-year. I told him she had never laid an egg. Her abdomen was six fingers, she was in good condition, and her pelvic bones were 1¼ inches thick.

CHART 6.

Six-finger Abdomen.

NERVOUS TEMPERAMENT.
$^1/_{16}$ pelvic bone..............280 eggs
⅛ pelvic bone..............265 eggs
$^3/_{16}$ pelvic bone..............250 eggs
¼ pelvic bone..............235 eggs
$^5/_{16}$ pelvic bone..............220 eggs

SANGUINE TEMPERAMENT.
⅜ pelvic bone..............205 eggs
$^7/_{16}$ pelvic bone..............190 eggs
½ pelvic bone..............175 eggs
$^9/_{16}$ pelvic bone..............160 eggs
⅝ pelvic bone..............145 eggs

BILIOUS TEMPERAMENT.
$^{11}/_{16}$ pelvic bone..............130 eggs
¾ pelvic bone..............115 eggs
$^{13}/_{16}$ pelvic bone..............100 eggs
⅞ pelvic bone.............. 85 eggs
$^{15}/_{16}$ pelvic bone.............. 70 eggs

LYMPHATIC TEMPERAMENT.
1-in. pelvic bone..............55 eggs
$1^1/_{16}$ pelvic bone..............40 eggs
1⅛ pelvic bone..............25 eggs
$1^3/_{16}$ pelvic bone..............10 eggs
1¼ pelvic bone.............. 0 eggs

He cautioned me to be careful, as he had always trap-nested his hens, and his record showed how many eggs they had laid. I replied, "If that is the case, her record shows that she has never laid an egg." He said no more then, but brought me another hen, asking me how many she would lay. I examined her for capacity. I found she was a six-finger-abdomen hen; her condition was good; her pelvic bones were $^1/_{16}$ of an inch thick; they were both alike as to thickness. I questioned him as to how he had fed her, and if she had been sick her first laying year. As he is one of the best breeders in the United States, I could depend on him knowing what he was talking about. I asked him then to take off his hat. I could see by the shape of his head he was a strictly honest man. I then told him that I had never raised that breed of hens, but if it was a Leghorn, it might lay 280 eggs its first year, and if a Plymouth

Rock, it might lay 270. He said her trap-nested record showed she laid 276 eggs from the time she commenced to lay in her pullet year until she had laid one year. "That's all right," I replied; "but what about the first hen we examined?" "We have never found any in the trap-nest from her," he said, "but she might be in the habit of laying in the yard." And as he was offered $1,000 for her, he was very anxious to get some chickens from her. I explained to him that while most typical beef hens could be made to lay a very small number of eggs in the spring when the crows laid, by feeding them a little lean meat and shrunken wheat and bran on a grass plot of white clover (if the blossoms of the white clover are clipped off), that his hen could not be made to lay, as she was a barren hen, as indicated by the rigid cord that connected both of the pelvic bones together, thus indicating that Nature never intended her to lay. I could name a number of professors and physicians that have told me they have discovered the same condition after they had taken my lessons.

The reader will please bear in mind that the two pelvic bones of a hen are not always of the same thickness. Some hens may have one pelvic bone thicker than the other; when this is the case, add the two together and half of the number will be the right thickness to judge by. For instance, if one pelvic bone was $\frac{1}{8}$ of an inch and the other one was $\frac{1}{4}$ of an inch, the added thickness would be $\frac{3}{8}$ of an inch; dividing this would give you $\frac{3}{16}$ of an inch as the thickness of one pelvic bone. Where one bone is thicker than the other, the thinnest one is on the left side of the hen.

Our next hen may be another six-finger-abdomen hen, as in Fig. 17, she may be in good condition, as in Fig. 19, and her pelvic bones may be $\frac{1}{8}$ of an inch thick, as in Fig. 25; she would be a 265-egg type hen.

Our next hen may be a six-finger-abdomen hen in good condition; pelvic bones $\frac{3}{8}$ inch; she would read six fingers abdomen; good condition; pelvic bones $\frac{3}{8}$ of an inch. By consulting Chart 6, we will find this is a 205-egg type hen.

Our next hen may be a six-finger-abdomen hen, in good condition; $\frac{1}{2}$-inch pelvic bones; this hen will be a 175-egg type hen.

Our next hen may be a six-finger-abdomen hen, in good condition; pelvic bones 1 inch. We look on Chart 6, and find that 1-inch pelvic bones indicate the 55-egg type hen.

Our next hen may be a four-finger-abdomen hen; she may be two fingers out of condition, as in Fig. 21, and her pelvic bones may be $1/16$ of an inch thick. We would read her as four fingers abdomen; two fingers out of condition; this would make her a six-finger-abdomen hen if in condition. We look on Chart 6 at $1/16$-inch pelvic bone, and find our last hen is a 250-egg type hen, if in condition, and it is up to us to put her in condition and keep her there as nearly as possible.

I will admit it is a hard proposition to keep the non-setting typical-egg type hen in condition, but the man that comes the nearest doing so is the best feeder. I will have more to say in regard to the matter of condition in the chapter on Judging Utility Fowls at the Poultry Shows. This work is a matter of line upon line, and I must necessarily repeat the same matter in some respects time after time. But as this is an educational more than an entertaining proposition, I hope that my readers will bear with me.

As I have said before, there are three types of hens. The hen listed on Chart 1 as $1/16$-inch pelvic bone is a typical egg-type hen, because all she consumes over bodily maintenance goes to the production of eggs. The hen listed as $3/8$-inch pelvic bone is a dual-purpose hen; half of her vitality is used in producing eggs and half in producing meat. The hen listed as $5/8$-inch is a typical meat-type hen; all she consumes goes to the production of meat, except what she uses in bodily maintenance. The hen listed as $1/16$-inch pelvic bone on Chart 2 is a typical egg-type hen; the hen listed as $3/8$-inch pelvic bone on same chart is a dual-purpose hen; and the one listed as $3/4$-inch pelvic bones is a typical meat-type hen; the same rule follows in all the charts. All the hens listed as $1/16$-inch pelvic bone are typical egg-type hens, and they can't be made to pay as a meat proposition. The hens listed in the center of each chart are the dual-purpose hens; they can be used as an egg and as a meat proposition. The hens listed on the bottom of each chart are the meat-type hens. Nature has fitted them for the production of flesh, and there is no human energy that can change them to a paying egg proposition.

Between the above three distinct types there are combinations of each adjoining type. This allows sufficient latitude for the preference of each individual breeder. A person can breed the typical egg-type hen and cock bird with pelvic bones $1/16$ of an inch thick. If he thinks this type is too delicate, he can breed from the $\frac{5}{16}$-inch pelvic bone stock; this is my favorite type; the hen of this type is better able to withstand the vicissitudes of the poultry-yard than her finer-bred sisters. I will have more to say along this line in the chapter on Broilers. I think we have given sufficient examples in Chapters III., IV., V., VI. and VII. to enable the reader to examine a hen so he may be able to arrive at her approximate value for the purpose he wishes to use her for.

In a previous chapter we have said there is occasionally found a hen seven fingers abdomen. If the reader finds one, he can score her by Chart 6 and add 15 eggs to the number indicated. For instance, if the hen is in good condition and measures seven fingers abdomen and her pelvic bones are $\frac{3}{8}$ inch thick, Chart 6 would indicate she is a 205-egg type bird; we then add 15 eggs to the 205, which gives the hen 220-egg capacity. If she is five fingers abdomen and two fingers out of condition, we call her seven fingers abdomen, and proceed as above, which gives us the same results.

There are two other matters I wish to call the attention of the reader to in this place. One is, that I have found hens occasionally that laid a great deal better by the trap-nest than they scored by the Hogan test, but it was owing to a mistake made in measuring their abdomens, owing to the rear of the breast-bone turning up, sometimes almost an inch over normal shape, thus indicating a smaller abdomen than really was the case. The other matter is a more serious one—in fact, very serious in some flocks. It is the **bagging down of the abdomen over the rear of the breast-bone.** Every hen used in the breeding-pen should be examined for this defect, for if one of them is bred from, she is almost sure to transmit her weak ovarian system to her offspring. Some of these hens will make remarkable egg records for a year or so, then will never lay another egg; and again, the eggs are liable to be very infertile and more or less thin-shelled; and if you have great numbers of hens, you can hardly tell when these hens stop laying for good, unless you trap-nest them, as their pelvic bones do not close up as readily as hens in normal condition.

An ounce of prevention is worth a pound of cure in this case, as it is very easy to prevent all this trouble. I meet hundreds of the above hens in my visits to poultry plants, but never have a case in my yards. I examine all my pullets when about a year old for possible breeders. If a hen satisfies me a to **Capacity, Type, and Prepotency,** I then hold her as if I were testing her for capacity, **except** that I hold her by the **right leg only.** I then lay my hand on her breast, so that it (my hand) will conform to her shape, and draw it slowly along her breast-bone (or keel) from front to rear. When my hand reaches the rear, if I feel the slightest indication of her abdomen dropping the least bit below the rear of the breast-bone, I reject the hen as a breeder, and thereby save myself a world of trouble in the future.

CHAPTER IX.

Prepotency.

We will take up in this chapter Prepotency, the science of breeding poultry, so that we can breed with a definite knowedge of what we are doing, and not leave it to intuition or chance. It is an old saying that "like begets like"; this seems to be true in some cases, but seems not to be true in other cases. Students of human nature can readily see where it has apparently failed. Some children will resemble and act like one parent and some will resemble and act like the other parent; then again, some children will be like neither of the parents. Breeders of horses and cattle are well aware of the variations in offspring from the type and characteristics of sire and dam. It is more through persistency in breeding than the general knowledge of any scientific principle that we have succeeded in producing the grand types of animals we see at our State fairs. The breeding of poultry is no exception to the above rule. While some breeders have good success in breeding for the desired type of bird, whether for fancy, for eggs, or for flesh, others will have very poor success.

The purpose of this chapter is to explain to the breeder who has had poor success a method that will enable him to breed with the full understanding as to what he is doing. It is a well-known fact among the clothing trade that if a woolen manufacturer has a sample of cloth presented to him, he can manufacture thousands of yards that will be an exact duplicate of the sample. The same is true in other industries. But suppose the reader gives an order to one of our well-known poultry-breeders for 1,000 pullets, to be delivered at four months old; these pullets to be housed, fed, and cared for as the breeder designates, and to approximately lay a certain number of eggs their first laying year; how many breeders do you suppose could fill the order? Until a majority of them can do so the poultry industry will not be on a business basis, but will be more or less of a gamble.

I have said that seemingly like does not beget like in some cases. We will take, for instance, a hen that is five fingers abdomen, in good condition, ¼-inch pelvic bones. She will scale up as a 205-egg type hen. We will mate up a pen of these hens with a 205-egg type cockerel or cock bird; we raise 100 pullets from this mating and they may scale 175-egg type. We then say, "Like does not produce like." Here is where we make a mistake. In one sense we are right, in another we are wrong. Nature makes no mistakes. We have mated 205-egg type male and female, and we get as a result 175-egg-type product. That's as plain as the nose on one's face, and we throw up our hands in despair and say, "It's all luck and chance." Another party mates up the same type of birds and gets a lot of pullets that average 210 eggs their first laying year; still another party mates up the same type of birds and does not get a chick.

The reader may smile, but this is no dream. A number of such cases have come under my observation. One case was that of a professor in one of the Southern California public institutions. He had a pen of twelve Black Minorcas, headed by a splendid-looking cock bird; also a pen of twelve Andalusians. He said there was something peculiar about these hens, and he wanted to know if I could detect it. I tested all the Andalusians, and told him they should average 140 eggs their first laying year, and I would expect twelve eggs out of every thirteen to be fertile. After testing the Minorcas, I told him they would average about 160-egg type, but if they were mine, I would not set any of their eggs while they were mated to the present cock bird, because I would not expect them to hatch, and if they did hatch, they would be degenerates. He said, "This is the second season I have bred from the birds; I always get good hatches from the Andalusians; but, although I see the rooster serve the hens, I have never been able to hatch a chicken from the Minorca pen." I replied, "He serves the hens out of sympathy."

Another case was a Barred Rock hen, the only one a neighbor had in a small flock of Houdans. He called me one day, saying he had a remarkable pullet at his place, and he wanted me to call over and tell him how many eggs she would lay her first laying year. She had been laying two months, and he was keeping her record. I went with him, tested the hen, and told him she might lay 250 eggs, but I did not think that any of them would hatch. After her first laying year was up, he showed me her record. She had laid 258 eggs, and although he had a good Barred Rock cock bird with her, and had set a number of settings under hens, he failed to hatch a single chick. I could cite a great number of such cases.

In the first of these cases the fault was with the male bird; in the last case the fault was with the hen; in both cases the trouble was caused by a lack of prepotency (amativeness), and not through any defect in the anatomy of the birds. Everything in the universe is governed by certain immutable laws. If we understand these laws and can discover a way to control them, we may be able to use them to our advantage. Does the reader ever stop to consider these matters? What, in your opinion, is the greatest effort of Nature? The writer thinks it is the effort to reproduce the species in all their different forms of animate and inanimate life. If the case were otherwise, this earth would be barren of grass and shrubs, of flowers and fruits, and of every living, moving thing on land and in the sea. What a desolate old world this would be with only bare dirt and rocks and water! And when we consider what a wonderful thing life is, can we doubt that Nature has made some extraordinary provisions for controlling its inception? In the wild state the survival of the fittest prevented degeneracy of the species, but under domestication birds cannot follow their instincts; and their owners should be familiar with Nature's laws in order to be able to breed intelligently.

When the writer was twelve years of age he took up the study of human nature, and later had help from that great teacher, Professor O. S. Fowler. Years of practice in dissecting and in anatomy and in the study of the skulls of animals and birds gave me the opportunity to study the construction of the different skulls and classify them as to the known habits of the birds or animals under consideration. The knowledge gained in this way was of inestimable value in later research in the selection and breeding of poultry. I am positive that without this early training I never could have accomplished what I have.

After raising my first lot of Leghorns in 1869, I decided to dis pose of all breeds but the Leghorns and Light Brahmas. I said I would raise Leghorns for eggs and Brahmas for meat. Up to that time I had not paid much attention to the individual laying qualities of the birds. Experience had taught me that the Light Brahma, when fed right and of the right age, made a delicious table-fowl, and I was led to believe the Leghorns were all great layers. That was a good many years ago; and we have made wonderful discoveries and progress in science and the arts since that time. The reader can imagine my surprise when I found by experience that some of my Leghorns laid very few eggs and laid them only in the spring months; others laid large numbers and laid late in the fall and early winter. In those days we had no cold-storage plants, and while eggs were very cheap in the summer, they were very dear in the winter, and I decided to experiment with my Leghorns, with a view to getting more eggs in the winter. After a few years of study and experiment, I mated the best egg-type birds and from some pens got good results, from other pens not so good, and from still others very poor results. My previous studies in anatomy had enabled me to select the matings from birds that were all of the same type, and I expected to raise a lot of poultry that would be duplicates of their parents, as far as their egg-laying qualities were concerned. But after numerous experiments in mating the 180-egg type cock bird with 180-egg type hens, I found I could not depend on getting definite results.

Some are born rich, some are born handsome, and some are born lucky. The writer was born with none of these gifts, but with a combination of faculties that compelled to invention, to wander and toil and delve in the fields, the byways, and the mines of the mysterious. These researches, with the aid received by studying the pioneers in the same lines of investigation, led to the discovery, as the writer thinks, of the fundamental principle that underlies the reproduction of the species. After a number of matings that were more or less discouraging failures, I decided to look to the brain of the bird as the seat of the cause of a great many of the variations between the characteristics of the offspring and those of the parents. I had previously demonstrated by experiment that environment had an influence on the shaping of the skull of the birds. By focusing on this subject the skull-knowledge I had gained in the previous nine years, I was led to think that brain governed most of the functions of the body, and if so, why not the reproductive function? I reasoned that as I had mated up several pens of the same type of hens with the same type of male birds, and that as there was no difference in their temperaments, that the hens all **looked alike,** all **weighed alike,** and were all in the same condition—or, in other words, they were all in **perfect condition** (to be more explicit, the hens were three fingers abdomen, pelvic bone $1/16$ of an inch thick; all hens were in good condition; the cock birds were two-fingers abdomens, in normal condition, and pelvic bones $1/16$ of an inch thick; all hens were alike and all cock birds were alike, and all were about a year old); that there must be something apart from the anatomy and physiology of the hen that governed or in some measure controlled the reproductive functions. As I had exhausted all my resources in the above lines, I was very reluctantly obliged to enter a new field of research—the field of Phrenology. I killed the cock birds that had given us the best results, boiled their skulls until free of flesh, and found them as in No. 1, Fig. 35. The skulls of the cock birds that gave the next best results were like No. 2, Fig. 35, and the skulls of the cock birds that gave the poorest results were like No. 4, Fig. 35.

The arrows A, B, C, and D show the base of the brain. If A were continued upward, it would pass through the projection $1/4$ of an inch from the end; if B were continued, it would pass through the projection about $1/8$ of an inch from the end; while C would be at the extreme end of the projection, and D would pass outside the skull. The part of the skull where the arrows 1, 2, 3, 4 point contains the rear lobe of the brain, and examination will show that the development of this portion of the brain corresponds to the shape of the skull at this point.

And right here is where we were on the point of the second great secret in breeding that would verify the saying that "Like begets like." The first discovery was, that if we wished to raise pullets that would be good layers, we would have to mate good-laying hens with the same type of male bird, and not with the meat type—that is, the male birds would have to be of the same temperament, of the same anatomy, and of the same physiology as the hen. I found that if I had a hen that laid 180 eggs by the trap-nest, and if I wanted to raise a lot of pullets that would average 180 eggs, I could not depend on the trap-nest to aid me any farther than to tell me the

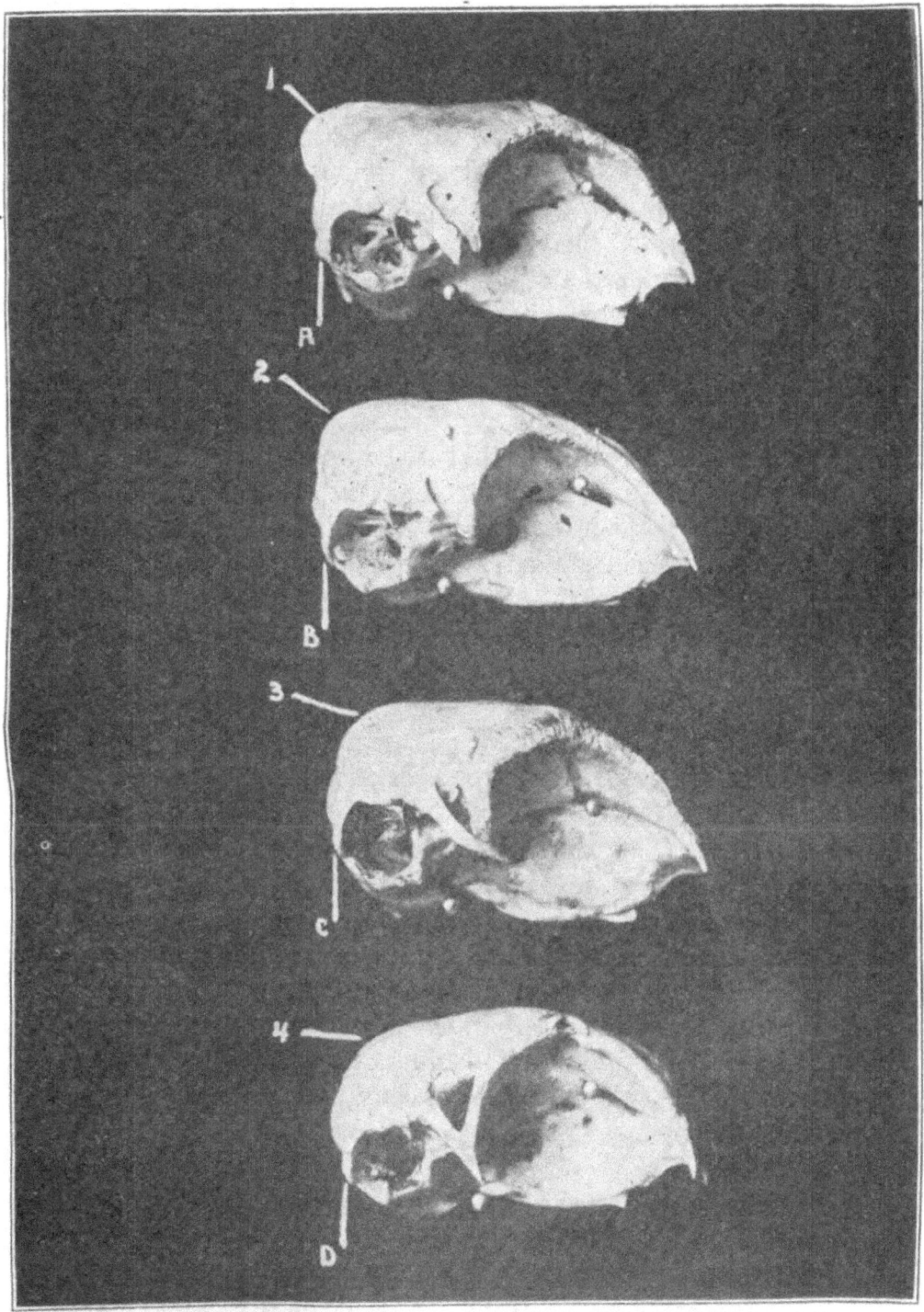

Fig. 35—Three degrees of amativeness (otherwise called "prepo-tency").

number of eggs a hen laid, what particular eggs she laid, and the progeny of each hen, both male and female. I also found great variations in type in the mature cockerels from each individual hen, which we considered was due to the difference in type of the male bird and the difference in vitality of one or both birds at different times during the breeding season; sometimes the hen, at other times the cock bird, transmitting their characteristics. When I was assured of this through numerous experiments, I reasoned that my failures were because **the male birds were of a different type from the hens,** and when I had demonstrated that the male birds **were** of a different physiology by practice and scientific measures, and mated accordingly, I flattered myself with the assurance that I had discovered all that was necessary in order to breed poultry intelligently. But after more experiments I was not wholly satisfied with results; and as I had adopted the motto, "Like begets like," I reasoned that although the birds we had mated were alike, as far as we could see, the remaining difference must be some place where I had failed to look for it. My knowledge of the different variations in form of the skulls of animals and birds **of the same breed,** together with the knowledge I possessed of human skulls, led me to investigate the head as the only remaining factor in the problem. When I reduced this proposition to a method, and when I was able to measure its potentiality, then I assembled the hens and cock birds, mating the 180-egg type hens and the 180-egg type cock birds, each bird with the same degree of prepotency. Then, and not until then, had I ever knowingly mated like to like. For years, like many others, I thought I had mated males to like females, but I was mistaken. And here is where I discovered my second great secret. After this I mated like to like more intelligently, and the results were more satisfactory.

I consider the selecting of the male birds for mating along anatomical and physiological lines, **together with the proper understanding and use of the faculty that governs the reproductive function,** as the greatest discoveries ever made in the poultry industry.

The reader may think there is very little difference in the skulls in Fig. 35. If you add an inch to the length of a man's legs, it does not seem to make much difference in his height, but if you add an inch to the end of his nose, it would make a great difference in his looks. I found this expansion on the back of the skull corresponded to the faculty of amativeness in the human family. I found that when it was large in both male and female the parents possessed the ability to transmit their **predominating characteristics** to their offspring. If the parents were fancy birds, their progeny would in some cases excel their parents in feather, vigor, and other good qualities. If the parents were of the egg type, some of the chicks would be as good and some better layers and more vigorous than the parents; if of the meat type, the progeny would be of a stronger constitution, of a quicker growth, and assimilate their food better—in a word, if both parents have this faculty (called "prepotency" by some) large, **the chicks will be more likely to be equal to, and some will excel, their parents along the lines in which the parents predominate.** If the parents have the faculty small, the chicks will not be so good as the parent stock, but will degenerate along the lines that the parents excel in. If a hen is a 200-egg type and she has this faculty small, she will be just as valuable as an egg-producer as if she had the faculty large, but she will be of no value as a breeder; she will be an old maid from choice, and her eggs will not be fertile, if she has the faculty small enough. If the male bird has it small, his eggs will not hatch well, and if small, they will not hatch at all. I have found a few cases where the cock bird had the faculty of prepotency (amativeness) large and failed to fertilize the eggs, but the cases were very rare, and I attribute it to weakened or diseased nerves; as, for instance, the nerves of the teeth or sciatic nerve in the human being.

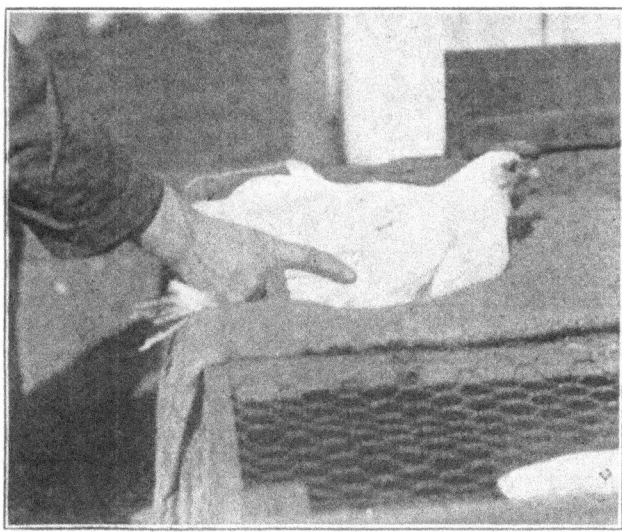

Fig. 36—Holding hen ready to put in sack.

Fig. 36 shows how to hold a hen before putting her in a sack to examine her for prepotency.

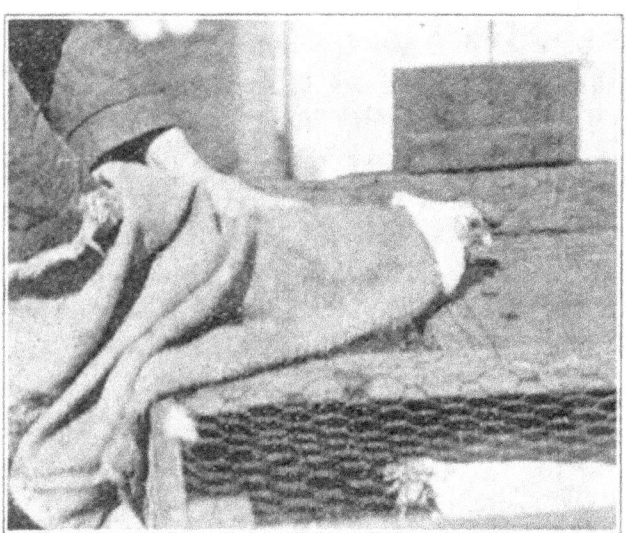

Fig. 37—Holding legs with right hand and gathering sack around legs with left hand.

Fig. 37 shows how to put her in the sack, holding legs with right hand, with back of hen against bottom of sack, and gathering sack around legs with left hand.

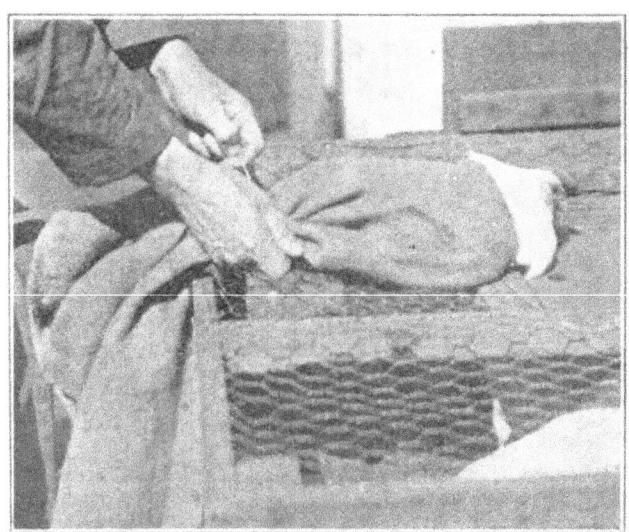

Fig. 38—Tying sack around legs so hen cannot move while examining her for prepotency. This method of holding the bird is only necessary while you are learning. If there is someone to hold the hen for you, it would be quicker. When you become skillful, you can hold the bird as in Fig. 43.

Fig. 38 shows tying sack around legs so that she cannot move while examining her for prepotency. (Cut a little off of the corner of the sack—just enough to get her head through, Hen in Fig. 38 is too far out of the sack.)

The best way for a beginner to learn how to handle a hen for prepotency is to select a hen you wish for the table. Cut the corner off of a gunny sack; hold her as in Fig. 36; put your hen in sack and tie her, as in Figs. 37 and 38; then make a hook of wire or a hair-pin, attach it to a string with small weight or stone; hang hen up against barn or shed, head down, back against building; take long-bladed pocket or other knife with sharp point, insert in hen's mouth, and draw across the roof of the mouth at the back of the brain at the junction of the neck, severing the blood-veins, then immediately force the knife through the roof of the mouth into the brain. The knife should be forced well into the brain, which will sever the nerves, and the bird will feel no pain; then insert hook in the nostril, and the weight will hold the neck straight, The hen should bleed freely. After bleeding has stopped, clean mouth and surrounding parts of blood, and place hen in some convenient place—on a box or coop. The thumb-nail on the left hand and nail on the forefinger of the right hand should be longer than the thumb and finger, so the flesh on end of thumb and finger will not prevent the nail from entering the slight depression between the skull and neck.

We will suppose the reader has handled the hen as suggested above. Lay the dead hen as in Fig. 39; take hold of comb or head and pull neck up with right hand, and while holding head up the neck will be stretched out. Turn the head down with right hand, so the back of the head will point up and beak will point down as much as possible. This will make the projection of the brain (arrow 1, Fig. 35) appear more prominent, so it will be easier to locate it; then draw ball of thumb of left hand down on head until you feel back of skull; when you feel back of skull with ball of thumb, then turn first joint of thumb down until thumb-nail fits in between end of skull and neck and well up against base of brain; then, while holding left hand and thumb as in Fig. 39, put forefinger of right hand at base of brain behind the ear, as in Fig. 39, between the neck and the skull and against the skull behind the ear, as in Fig. 39. The ear can readily be discovered by lifting up its hairy covering. The thumb-nail must be held perfectly straight across the neck, as in Fig. 39, and not sideways; and the forefinger must be held perfectly at right angles with the thumb, or the length of projection (arrow 1, Fig. 35) from the base of the brain (arrow 4, Fig. 35) cannot be measured accurately.

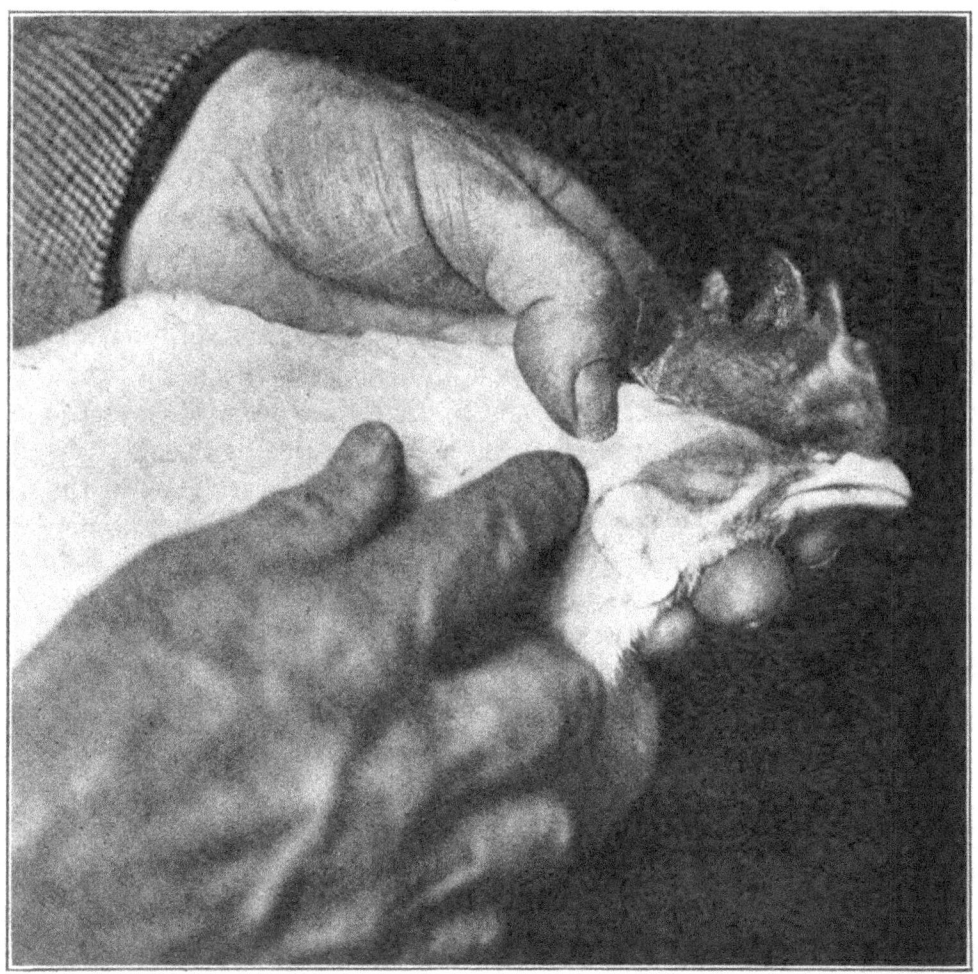

Fig. 39—Showing thumb ⅛ of an inch ahead of forefinger; indicating hen is totally lacking in prepotency. (See Skull No. 4, Fig. 35.)

The reader will notice that my thumb-nail is ahead of my forefinger-nail in Fig. 39; this indicates that this hen is wholly lacking in the ability to transmit any redeeming qualities to her offspring, also that she has no desire for offspring. If this were a male bird, the eggs from his matings would be infertile. Fig. 40 shows thumb on line with forefinger. Matings from this type of head would not produce very fertile eggs, and the progeny would deteriorate each year if they were bred from stock with heads like this. If the parents were 200-egg type, their egg-yield and vitality would be reduced each generation of breeding. If they were of the beef type, their vitality and ability to produce flesh economically would diminish with each generation. If they were a fancy type, the breeder would be up against a stone wall of discouraging experiments.

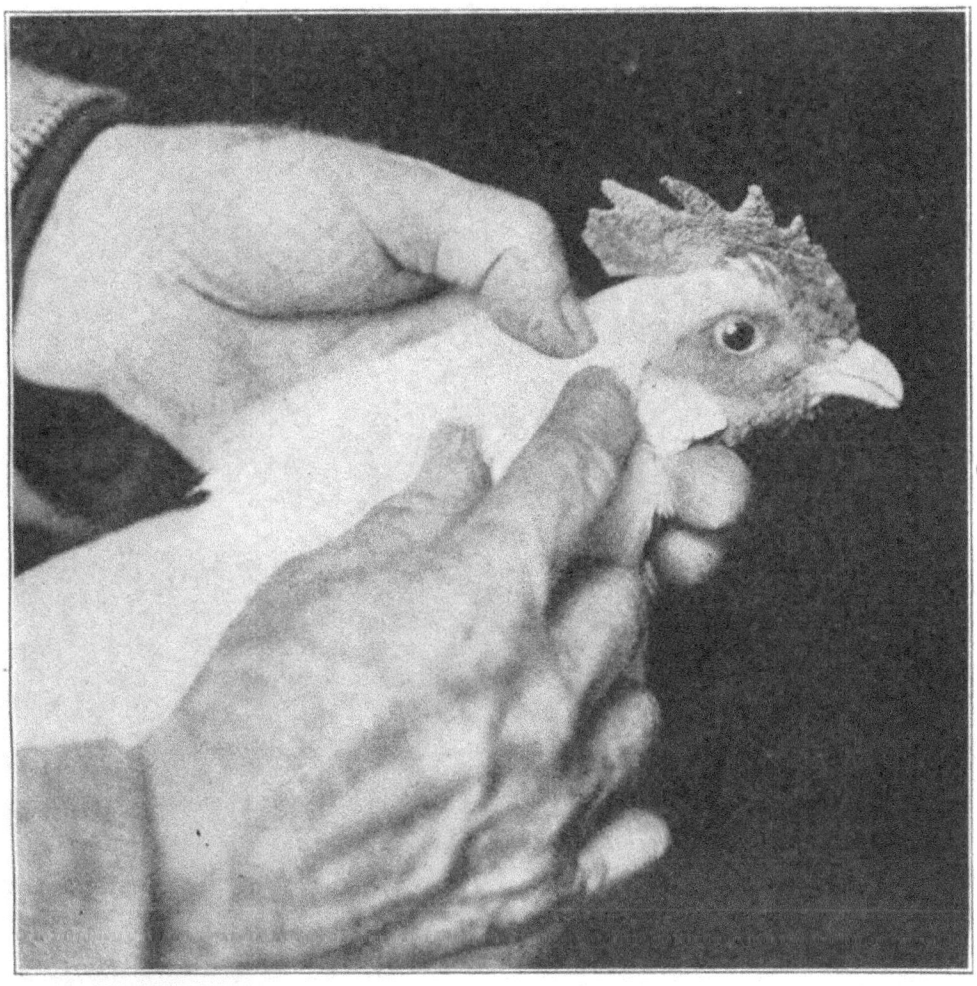

Fig. 40—Thumb even with forefinger; indicating she has prepotency small. (See Skull No. 3, Fig. 35.)

I would advise the reader to take special notice of Fig. 43, as this cut shows the method of determining prepotency plainer than any of the others.

Fig. 41 shows a hen with prepotency full—i. e., thumb ⅛ of an inch behind forefinger. Sometimes a poultryman will be lucky enough to mate up a lot of pens of the right type for his purpose with heads like Figs. 41, 42, and 43. His business prospers, and his neighbors call him "lucky." While others are going bankrupt raising poultry, he holds his own and is making a good living. Figs. 42 and 43 show a hen with an excellent head for breeding purposes. The thumb in this case is ¼ of an inch behind the forefinger. If this hen is mated to a male bird of the same type and prepotency, her eggs will be very fertile, and a large number of the progeny will be equal to and

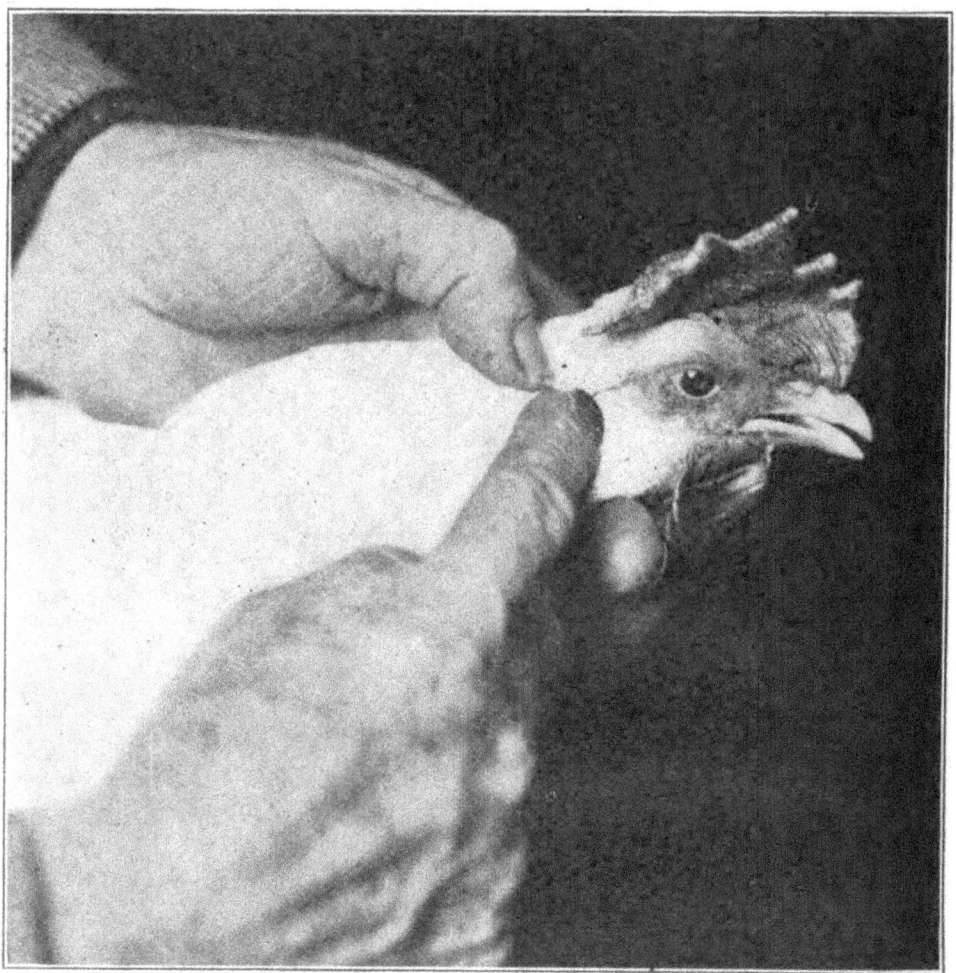

Fig. 41—Showing thumb 1/8 of an inch behind forefinger; indicating hen has prepotency full. (See Skull No. 2, Fig. 35.)

some will excel the parent stock in the lines that predominate in the parents. **By selecting these few specimens each season for breeding,** it is possible to breed a highly valuable type in the course of time. Fig. 43 shows how to hold a bird between the knees after you become proficient in testing the head while the bird is in a sack.

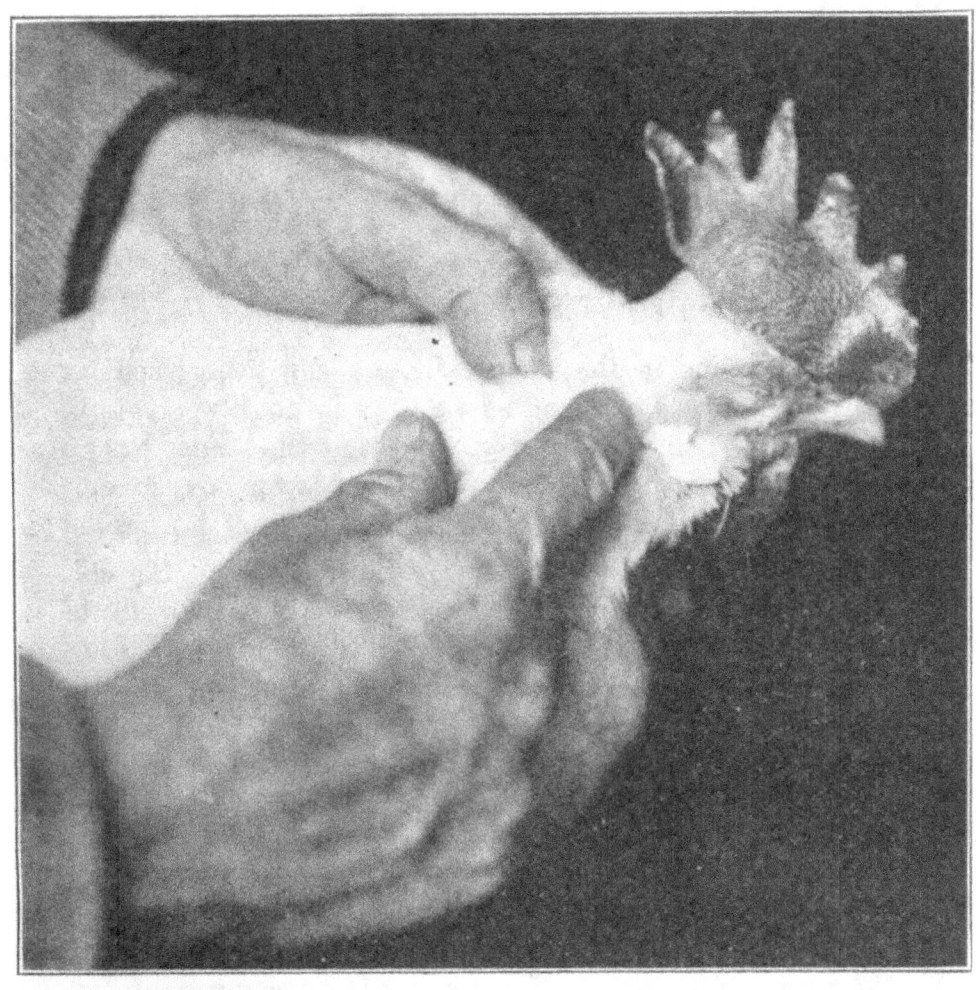

Fig. 42—Showing thumb ¼ of an inch behind forefinger; indicating hen has prepotency large. (See Skull No. 1, Fig. 35.)

CHAPTER X.

TESTING HENS ON A LARGE SCALE, USING CHARTS 44 AND 45.

I will describe in this chapter how I cull hens when we have large numbers of them, as we have in poultry plants in California. I shall take it for granted that the reader has no method of selecting the good from the poor layers, except, perhaps, the "Walter Hogan System" or some of its pirated forms that are now used extensively in all parts of the civilized world, and which is based on the theory that the value of a hen as an egg-producer depends on the relative distance apart of her pelvic bones and the thinness of same. We will suppose the reader has 300 hens; one lot are about a year and four months old, another lot are about two years and four months old, and another lot are about three years and four months old. Each lot has been kept in separate yards, so there can be no mistake in regard to their ages, or they have been toe-punched or otherwise marked. We notice more or less feathers lying around the yard, thus indicating the season of the year when moulting is near at hand. Everything else being equal, the poorest hen moults first, and if she is a **very poor layer,** she will stop laying when she begins to moult and will not lay again until the crows lay in the spring. We consider it is about time to cull out the poor layers and send them to market.

The next thing that comes to mind is the question, "What is a poor layer?" That all depends on the price you get for the eggs, the price of feed, houses, etc. I raised poultry in Todd County, Minnesota, in 1886 and 1887, and sold good lumber at the saw-mill for $5.00 per 1,000 feet. Wheat was about 1 cent per pound, and wheat screenings for chicken feed could be had for the hauling. It is very evident that a poorer class of layers might have been kept at a greater profit when supplies were at that low price than can be profitably kept when supplies are as high-priced as they are at the present time of writing (June, 1913). So the reader can see that the matter of the profitable hen is a local matter. At this writing you can buy nearly two bushels of wheat in some parts of Minnesota for what you will pay for one in California. I was told a few days ago that you could buy twice as much oats at the present time

THE CALL OF THE HEN. 83

in Minnesota as you can in California for the same money.

When studying Charts 44 and 45 we see there are certain figures lined off from the rest; this is for the purpose of aiding the reader at a certain time each year to select the poor layers from the good ones without using the charts, thereby saving the time necessary to look over the chart and classify each hen.

Charts 1, 2, 3, 4, 5, and 6, as the reader will learn by bearing in mind the following instructions, need be used only to determine the laying-score of the individual hen.

One-finger Abdomen.	Two-finger Abdomen.	Three-finger Abdomen.
1/16 pelvic bone... 36 eggs	1/16 pelvic bone... 96 eggs	1/16 pelvic bone... 180 eggs
1/8 pelvic bone... 32 eggs	1/8 pelvic bone... 87 eggs	1/8 pelvic bone... 166 eggs
3/16 pelvic bone... 28 eggs	3/16 pelvic bone... 78 eggs	3/16 pelvic bone... 152 eggs
1/4 pelvic bone... 24 eggs	1/4 pelvic bone... 69 eggs	1/4 pelvic bone... 138 eggs
5/16 pelvic bone... 20 eggs	5/16 pelvic bone... 60 eggs	5/16 pelvic bone... 124 eggs
3/8 pelvic bone... 16 eggs	3/8 pelvic bone... 51 eggs	3/8 pelvic bone... 110 eggs
7/16 pelvic bone... 12 eggs	7/16 pelvic bone... 42 eggs	7/16 pelvic bone... 96 eggs
1/2 pelvic bone... 8 eggs	1/2 pelvic bone... 33 eggs	1/2 pelvic bone... 82 eggs
9/16 pelvic bone... 4 eggs	9/16 pelvic bone... 24 eggs	9/16 pelvic bone... 68 eggs
5/8 pelvic bone... 0 eggs	5/8 pelvic bone... 15 eggs	5/8 pelvic bone... 54 eggs
	11/16 pelvic bone... 6 eggs	11/16 pelvic bone... 40 eggs
	3/4 pelvic bone... 0 eggs	3/4 pelvic bone... 26 eggs
		13/16 pelvic bone... 12 eggs
		7/8 pelvic bone... 0 eggs

FIG. 44—Showing first chart for commercial egg plant where large numbers of hens are kept.

All hens with one- and two-finger abdomens are sold at the end of the first laying season, and three-finger hens, with pelvic bones 3/8 inch thick or more, are likewise disposed of. In California we do not figure that a hen that lays less than 10 dozen of eggs in her first laying season will pay to keep into the second (always taking into consideration the condition of the hen at the time of examination.)

The first figures underlined in Chart 44 are in the column indicating three fingers abdomen, from $1/16$-inch pelvic bone to $5/16$-inch pelvic bone. The second are in the column indicating four fingers abdomen, from $1/16$-inch pelvic bone to $7/16$-inch pelvic bone. The third are five fingers abdomen, from $1/16$-inch pelvic bone to $9/16$-inch pelvic bone. The fourth are six fingers abdomen, from $1/16$-inch pelvic bone to $11/16$-inch pelvic bone.

We will make a copy of Charts 44 and 45 on a piece of white cardboard and hang it up in a convenient place in

Four-finger Abdomen.	Five-finger Abdomen.	Six-finger Abdomen.
$1/16$ pelvic bone......220 eggs	$1/16$ pelvic bone......250 eggs	$1/16$ pelvic bone......280 eggs
$1/8$ pelvic bone......205 eggs	$1/8$ pelvic bone......235 eggs	$1/8$ pelvic bone......265 eggs
$3/16$ pelvic bone......190 eggs	$3/16$ pelvic bone......220 eggs	$3/16$ pelvic bone......250 eggs
$1/4$ pelvic bone......175 eggs	$1/4$ pelvic bone......205 eggs	$1/4$ pelvic bone......235 eggs
$5/16$ pelvic bone......160 eggs	$5/16$ pelvic bone......190 eggs	$5/16$ pelvic bone......220 eggs
$3/8$ pelvic bone......145 eggs	$3/8$ pelvic bone......175 eggs	$3/8$ pelvic bone......205 eggs
$7/16$ pelvic bone......130 eggs	$7/16$ pelvic bone......160 eggs	$7/16$ pelvic bone......190 eggs
$1/2$ pelvic bone......115 eggs	$1/2$ pelvic bone......145 eggs	$1/2$ pelvic bone......175 eggs
$9/16$ pelvic bone......100 eggs	$9/16$ pelvic bone......130 eggs	$9/16$ pelvic bone......160 eggs
$5/8$ pelvic bone......85 eggs	$5/8$ pelvic bone......115 eggs	$5/8$ pelvic bone......145 eggs
$11/16$ pelvic bone......70 eggs	$11/16$ pelvic bone......100 eggs	$11/16$ pelvic bone......130 eggs
$3/4$ pelvic bone......55 eggs	$3/4$ pelvic bone......85 eggs	$3/4$ pelvic bone......115 eggs
$13/16$ pelvic bone......40 eggs	$13/16$ pelvic bone......70 eggs	$13/16$ pelvic bone......100 eggs
$7/8$ pelvic bone......25 eggs	$7/8$ pelvic bone......55 eggs	$7/8$ pelvic bone......85 eggs
$15/16$ pelvic bone......10 eggs	$15/16$ pelvic bone......40 eggs	$15/16$ pelvic bone......70 eggs
1-in. pelvic bone......0 eggs	1-in. pelvic bone......25 eggs	1-in. pelvic bone......55 eggs
	$1 1/16$ pelvic bone......10 eggs	$1 1/16$ pelvic bone......40 eggs
	$1 1/8$ pelvic bone......0 eggs	$1 1/8$ pelvic bone......25 eggs
		$1 3/16$ pelvic bone......10 eggs
		$1 1/4$ pelvic bone......0 eggs

Fig. 45—Showing second chart for commercial plant where large numbers of hens are kept. Four-finger hens with pelvic bones more than $7/16$ inch thick, five-finger hens with pelvic bones more than $9/16$ inch thick, and six-finger hens with pelvic bones more than $9/16$ inch thick are sold at the end of the first laying season, because they will not pay us in California for more than one laying season.

the yard where the sixteen-months-old hens are penned. We will suppose that the hens are all closed in the house or houses. We put catching-coop in position as in Fig. 2, and drive hens in same as in Fig. 1. When there are enough hens in the coop, shut down slide-door that holds them in. In this case it is necessary to keep only four fingers in mind; any four you prefer will do.

Here in California we use the figures 5, 7, 9, and 11 for the hen sixteen months old; this means a three-finger-abdomen hen, $5/16$-inch pelvic bone; four-finger-abdomen hen, $7/16$-inch pelvic bone; five-finger-abdomen hen, $9/16$-inch pelvic bone; six-finger-abdomen hen, $11/16$-inch pelvic bone. Anything below this line (that is, any hen having a thicker bone in the different classes) goes to market. For the twenty-eight-months-old hens we assume that they are hatched in March and sold in the summer. We use the figures 3, 5, 7, and 9 for the three-, four-, five- and six-finger-abdomen hens. For the forty-months-old hens we use the figures 1, 3, 5, and 7 for the three-, four-, five- and six-finger-abdomen hens. You perceive that the older the hen the greater the number of eggs she must have laid in her first year. Here in California we keep large numbers of hens, and in this way we can sort out the market here each year in a short time, as we do not have to stop and figure out the percentage of loss for each year of age, as these figures come **near enough** to suit our purpose. If they do not suit the local market, the reader can use any figures that will.

I shall give a few examples only to show how we would proceed to cull out the hens. The reader must be familiar with the general principles of capacity, condition, and type. He should by this time have familiarized himself with the charts. Now, if he prepares the figures as I have directed, he will experience no difficulty in determining in a moment just where and what to do with each individual hen. We establish a certain standard of production for the first laying season, in order to know how long to keep her. You may take 80 eggs for one season, 120 eggs for two seasons, and 150 eggs for three seasons, or any other set that suits your local conditions. Here we take about 120, 150, and 180 eggs as the standard; that is, a hen must be able to lay about this number in her first laying year in order to stay with us for two, three, or four seasons. With this explanation, we shall proceed to cull, putting into the shipping-coop all hens that fall below our standard, and dropping in the yard where we stand any that we desire to keep.

Now, take a hen out of the catching-coop as in Fig. 3, and hold her as near as possible as in Fig. 5. Place hand on abdomen. She may be one finger abdomen, in good condition; her pelvic bone may be $1/16$ of an inch thick; her capacity is three dozen eggs her first laying-year. She has laid all these eggs and will lay no more until the next spring when the crows lay, and eggs are cheap; so we decide to put this hen in the shipping-crate, to be sent to market.

We take another hen from the catching-coop, and go through the same process. She may be a two-finger-abdomen hen, in good condition; her pelvic bones $1/16$ of an inch thick; this indicates a hen that may lay eight dozen of eggs her first laying-year. As a rule, when hens are so fed and cared for, they will lay their maximum number of eggs their first laying year; they will, as a rule, lay **about** 15 per cent less each year after, provided they are given the same care and feed. In this case the hen in hand might lay about 85 eggs; if you think that will pay you, let the hen drop out of your hands into the yard where you are standing; if you think it will not pay to keep her, put her in the shipping-crate for the market.

The next hen may be two fingers abdomen, one finger out of condition, as in Fig. 20, with pelvic bones $1/4$ of an inch thick. If this hen's comb and wattles are red and the hen is strong and active, being one finger out of condition indicates that she is not being properly cared for, either in food or environment, or both; in the condition she is in at present, if continued the whole year, she might lay about 69 eggs, while if kept in normal condition, she might lay 138 eggs. (See Chart 3.) So we will call her a good hen, and drop her.

The next hen may be three fingers abdomen, $5/16$-inch pelvic bone, and in normal condition. If this hen were in Petaluma, we would drop her, as she would be a paying hen. By referring to the chart, you will see that she is a 124-egg type hen. You must bear in mind constantly that a thick bone in a hen of small abdominal capacity would mean a practical nonproducer, while the same thickness of bone in a hen of much larger capacity would mean simply a more beefy hen.

The next hen may be three fingers abdomen, in normal condition, as in Fig. 19, and pelvic bone $3/8$ of an inch thick.

This hen has the same abdominal depth as the preceding, but her pelvic bones being $3/8$ of an inch thick would make her a 110-egg type hen, and with us no hen that lays 120 eggs pays to keep two seasons. We put this hen in the shipping-crate for market, as it will not pay to keep her any longer, if in Petaluma. She will not pay for her board after this time and leave enough profit.

The next hen may be four fingers abdomen, in normal condition, and $7/16$-inch pelvic bone. She, being a 130-egg type hen, it will pay to keep her another year, so we drop her.

The next hen may be four fingers abdomen, in normal condition, and $1/2$-inch pelvic bones; this hen will lay approximately 115 eggs her first laying year, but not enough her second year; so we put her in the shipping-crate for market.

The next hen may be a five-finger-abdomen hen and in good condition; $9/16$-inch pelvic bone. She is a 130-egg type hen, so we drop her. While this hen has a pelvic bone $9/16$ of an inch thick, she has the abdominal capacity to supply herself with food enough to lay a profitable number of eggs and put on flesh at the same time.

The next hen is five fingers abdomen, in normal condition, and $5/8$-inch pelvic bones; this is a 115-egg type hen, so we put her in the shipping-crate. The hen we had just before this one was kept; but when we come to the $5/8$-inch pelvic bone, we decide that we have reached the lowest limit of egg-production.

The next hen may be six fingers abdomen, in normal condition, and $11/16$-inch pelvic bone; she will be a 130-egg type hen, so we drop her.

The next hen may be six fingers abdomen, in normal condition; pelvic bones $3/4$ of an inch thick; she will be a 115-egg type hen, so we will put her in the shipping-crate.

The next hen may be three fingers abdomen, three fingers out of condition, and $1/8$-inch pelvic bones. If her comb and wattles are pale and bloodless, she is no doubt diseased and should be disposed of; but if her comb and wattles are red, it indicates, as a rule, that she is out of condition on account of accident or lack of feed. In her present condition she scores 166-egg type. If we get her in one finger better condition, she will measure four fingers abdomen, and score 205-egg type; if we can get her in two fingers better condition, she will measure five fingers abdomen and may be $3/16$-inch pelvic bones, on account of becoming a little more fleshy, and score 220-egg type; and if we can get her in three fingers better condition, she would then be in normal condition, and her pelvic bones might be $3/16$ or $1/4$ inch thick; if the latter, she would score 235-egg type. (We will have more to say on the changing of thickness of the pelvic bone in the last of Chapter XVIII.)

We will continue selecting or separating the good from the poor layers in the same manner, keeping every hen for another year in the three-finger-abdomen class that is $5/16$-inch pelvic bone and thinner, and sending every hen to market that is over $5/16$-inch pelvic bone in the three-finger-abdomen class; keeping every hen in the four-finger-abdomen class that is $7/16$-inch pelvic bone and thinner, and sending every hen to market that is over $7/16$-inch pelvic bone in the four-finger-abdomen class; keeping every hen in the five-finger-abdomen class that is $9/16$-inch pelvic bone and thinner, and sending every hen to market that is over $9/16$-inch pelvic bone; keeping every hen in the six-finger-abdomen class that is $11/16$-inch pelvic bone and thinner, and sending every hen to market that is over $11/16$-inch pelvic bone thick.

I want to say here that there is nothing arbitrary in regard to Charts 44 and 45. Each poultryman can draw the lines where he thinks it will best suit his purpose. A great many years of experimenting has led the writer to believe these charts answer the purpose very well.

We have disposed of all the one-year-and-four-months-old hens, and will move our outfit to the two-year-and-four-months-old hens, and arrange the catching-coop and charts as in the first case.

The first hen we take from the coop may be a one-finger-abdomen hen, in good condition. All one- and two-finger-abdomen hens in good condition over one year and four months old, as a rule, should be disposed of. There is no profit in them after they have laid their allotted number of eggs their first season—or, in other words, after they commence to moult in their first laying year; so after this we will not consider them in this connection.

There is a great difference in the number of eggs a flock of hens will lay each year as they grow older. Some will lose 5 per cent, some 10 per cent, some 15 per cent, and some 20 per cent. Some will not lay anything (this will be explained later) after their first laying year. It depends altogether on the **vitality** of the hen and how she has been fed and raised; and the variations in the percentage of eggs laid by exactly the same type of hens will vary with different poultry-keepers and also with the same poultry-keeper, varying more or less in each separate pen, proving that environment has more or less to do with egg-production, all other things, as far as human knowledge is concerned, being equal. Some people who are good mathematicians, but who are wholly ignorant of animal nature, look surprised when I explain to them the difference between classifying the production of a number of like machines with the production of a number of hens of the same score in egg-production. As a scientific proposition, it is impossible to write a chart beforehand that will fit every case. If we took 1,000 hens of any pronounced type—say 100-egg type, which were fed, housed, and cared for in exactly the same manner, and one of them laid 5, 10, or 15 eggs more or less some year than the other 999 hens, it would prove our contention or theory, from a scientific point of view. I am sure that 100 expert poultrymen could take 100 hens of the same general type that would score the same egg-capacity and would all be in the same condition, and each poultryman feed and care for his 100 birds for four years the best he knew how, and very few of them would agree on a set of figures that would give the percentage of decrease in egg-production each year. The one who fed the heaviest and produced the most eggs would have the largest percentage of decrease, while the ones who bred for hatching eggs and did not force their hens with condiments and stimulants would get the least number of eggs and the lowest percentage of decrease, not figuring the percentage of decrease from the number of eggs actually laid, but from what the hen would lay each year.

The writer does not claim that he has discovered a system that will infallibly give results just as he has written them. No poultryman needs to be told this, but for the benefit of the amateurs I have inserted the above caution. The writer claims, by years of investigation and practice, to have formulated a poultry code as contained in this book that is commercially the approximation of perfection.

We will return to our two-year-old hens. We said all one- and two-finger-abdomen hens should be sold and we will consider them no more than to put them in the market crates when we find one. The reader will remember that in selecting the sixteen-months-old hens we retained only those in the three-, four-, five-, and six-finger-abdomen columns that measured $5/16$, $7/16$, $9/16$, and $11/16$ of an inch or less, and everything below these lines went to market. In the show-room, when the writer judges utility birds, we use the charts, so as to score each bird according to its capacity for egg-production; but when we cull the poultry on commercial plants, in order to save the time of looking on the charts, we keep in mind only four figures for the hens of any age that we are examining. For hens about sixteen months old, we use the figures 5, 7, 9, and 11; for hens with three-finger abdomens, we use the figures $5/16$; for four-finger-abdomen hens, $7/16$; for five-finger-abdomen hens, $9/16$; and for six-finger-abdomen hens, $11/16$. All under three fingers abdomen go to the market and all under the line go also.

For the two-year-and-four-months-old hens we keep in mind the following figures: 3, 5, 7, and 9. For the three-finger-abdomen hen, $3/16$-inch pelvic bone; four-finger-abdomen hen, $5/16$-inch pelvic bone; five-finger-abdomen hen, $7/16$-inch pelvic bone. Everything below these figures goes to the market; also all one- and two-finger-abdomen birds there may be in the lot.

We now go to the hens that are three years and four months old. Any one- and two-finger-abdomen birds that we may find go to market and all the three-finger-abdomen birds below $1/16$-inch pelvic bones. For the three-year-and-four-months-old birds we bear in mind 1, 3, 5, and 7. Three-finger-abdomen hen, $1/16$-inch pelvic bones; four-finger-abdomen hen, $3/16$-inch pelvic bones; five-finger-abdomen hen, $5/16$-inch pelvic bones; and six-finger-abdomen hen, $7/16$-inch pelvic bones. All below these lines go to market.

If the reader has some good hens that he wishes to breed from, he can use the figures. 1, 3, and 5.

The fourth year, when he wishes to select from the four,- five-, and six-finger abdomen hens, it will be: Four-finger-abdomen hen, $1/16$-inch pelvic bones; five-finger-abdomen hen, $3/16$-inch pelvic bones; and six-finger abdomen-hen, $5/16$-inch pelvic bones. Very few will want to keep hens as long as this. They will be five years and about four months old when you will sell them. Most people here sell them about the time they commence to moult—after they are two years old; but I selected two hens used at the California State Poultry Experiment Station to test this method as far as the egg-laying qualities were concerned, and the hens I selected as hens that would pay at four years made a good paying record.

The reader will understand that the way we have just been selecting the paying hens is the way we select when we have large numbers; this is the way I selected 1,600 hens in six hours at the poultry farm of the Ukiah State Hospital, Mendocino County, California, and at other State hospitals and poultry plants. We do not have to stop to figure out the percentage of loss of each bird. You can take any combination of figures you wish, as ¼-inch, ⅜-inch, ½-inch, ⅝-inch, for sixteen months-old birds; $1/16$-inch, $3/16$-inch, $5/16$-inch, $7/16$-inch, for twenty-eight-months-old birds. You can figure out the percentage of loss each year and take a combination of figures that will suit your purpose. You have only to carry four figures in your mind. The percentage of loss each year is computed by good poultrymen to be from 10 to 20 per cent in egg-production on plants that are run for hatching eggs. If you force your hens with an excess of meat and condiments, the loss will be according to how you feed them, and no one can tell what it may be but yourself. Some poultrymen will get all there is in a hen out of her the first season, then sell her.

CHAPTER XI.

The Male Bird.

This is not a treatise on cattle or horses, but we have to use them very often to illustrate the matter in hand. Stock-raising has been brought to more of a science than poultry-raising, and is well understood by thousands of our progressive farmers. I have met hundreds of them who could describe to me the points I would have to consider in selecting a good-paying butter-fat, beef or milk proposition, both in dam and sire; and while there may be as many poultrymen who understand the selection of poultry, both male and female, for egg- and meat-production, I have failed to meet them; and while I was made the butt of ridicule by the poultrymen when I issued my first pamphlet, entitled the "Walter Hogan System," in March, 1905, the stock-raisers who were interested in poultry stood by me to a man. The reason was, that the cattlemen had been studying along the utility lines in both sire and dam in order to develop the milk, butter-fat, and beef-producing capacities of their cattle. It was a comparatively easy proposition for them. The form of the animals was plainly to be seen. They were not covered with a coat of fluff and feathers that hid the shape and form of the subject. It was easy to distinguish between the cat ham of the butter-fat type and the full, deep ham of the beef type. It was no trouble to compare the udders, milk-veins, and wedge-shape type of the Jersey with the full, rounded build of the Hereford or Polled Angus.

On the other hand, the poultrymen, to some extent, were deceived by the appearance of their hens. Take, for instance, the Cochin and the Bantam; they would hold about the same relation to each other as the lordly Durham would to the fine-bred Devon, yet I have found Bantam hens with as deep abdomen as a great Cochin hen; and it is my opinion that if poultry were as bare of feathers as cattle are, the poultry industry would be as far advanced at present as is the cattle business.

The greatest impediment to the successful breeder of poultry has been the inability to select the male bird of the required type. The custom in vogue at the present writing with most poultrymen is to trap-nest their hens and raise cockerels from the best layers as indicated by the trap-nest. The trouble with this method is, that while the hen may lay a large number of eggs, she may not have the faculty to transmit her laying qualities to her offspring, and her cockerels may be deficient in both egg-laying qualities and the ability to transmit what good qualities they may possess to their progeny.

Again, I have seen a great many cases where poultry-farmers would send away and buy a lot of cockerels. The man that raised and sold them had no knowledge of how to classify them, and the man who bought them knew he was buying cockerels and that is all he did know about them. He could not be sure whether they would increase his egg-yield or not. He had to pay his money and take chances. It was nothing more or less than a gamble; but the days of gambling in the poultry business are passed for the intelligent, progressive poultryman, no longer will he be obliged to trust to luck or intuition. He will be able to select his male birds with as much assurance as his hens, and instead of groping in the dark, he will have the satisfaction of seeing and knowing just what he is doing by bearing in mind the instructions in this chapter.

The reader will by this time be familiar with the different types and capacities of hens, and will not be surprised to learn there is a similar number of variations in the male birds; and if one wishes to produce a certain type and capacity in a pullet or cockerel, he must select the parent birds that will produce that type. We know how to select the hen; we will now take up the study of how to select the male bird.

We go through the same movements in selecting or testing the male bird as we do in selecting the hen, but we use a different set of charts. For example, it is possible for a hen to change from six to three fingers in abdominal capacity within a month and be healthy and active, and in another month to return back to her original six-finger capacity, but it is not so with the male bird after he is mature. I have tested male birds at nine months of age that scored four fingers abdomen, $1/16$-inch pelvic bone, that did not change for four years, except that, their pelvic bones being $1/16$ of an inch thick at nine months old, I have found them to be $1/8$ of an inch thick at eighteen months old. They had increased in thickness of bone from $1/16$ to $1/8$ inch. These were egg-type male birds; the meat-type will vary more or less in the thickness of the pelvic bones—depending on how much flesh they put on or lose between the different times of examining them.

It will be easy to distinguish the egg-type cock bird from the meat-type bird; the former has thin pelvic bones, **whether in flesh or not,** while the latter has thick pelvic bones with a more or less lump of gristle on the end of them, whether he is thin or in good flesh. I have found that in classifying the male bird as we have the hen as to type and capacity for a certain egg-yield it requires less abdominal capacity in the male bird than in the female. For instance, the male bird that is two fingers abdomen and $1/16$ of an inch pelvic bone is the same type and capacity for breeding purposes as the three-finger-abdomen hen, $1/16$-inch pelvic one. The male of the same class, as regards capacity, does not require as large an abdomen as the female; this is so self-evident that it would be a waste of time to try to explain the reason for it.

I have heard poultrymen say that the male bird is half of the flock. I wonder if they stop to consider whether this is so or not. My birds are wonderful layers, and I mate one male bird to every twelve hens, and from a breeder's point of view I consider my male birds a great deal more than half the flock.

If I mate 100-egg type cock birds with 200-egg hens, the progeny may lay about 150 eggs, thus reducing my egg-yield about 25 per cent in the progeny of each of the twelve hens. For this reason I have given as much thought to the male bird as I have to the hen; and in arranging the charts for the male birds have experienced a great deal of difficulty, as it takes years of time and hundreds of matings to arrive at conclusions that would be approximately correct. In any one case, everything else (type, capacity, and breed) being equal, care and environment have a dominating influence on the product, whether eggs or meat; consequently, if a number of investigators were working on this proposition, using the same system of selection, they could not help but arrive at somewhat different conclusions as to figures, but that would not affect the value of the system.

MALE BIRD—CHART A.

One-finger Abdomen.

$1/16$ pelvic bone...........84-egg type
$1/8$ pelvic bone...........75-egg type
$3/16$ pelvic bone...........67-egg type
$1/4$ pelvic bone...........58-egg type
$5/16$ pelvic bone...........50-egg type
$3/8$ pelvic bone...........41-egg type
$7/16$ pelvic bone...........33-egg type
$1/2$ pelvic bone...........24-egg type
$9/16$ pelvic bone...........16-egg type
$5/8$ pelvic bone........... 7-egg type
$11/16$ pelvic bone........... 0 egg-type

MALE BIRD—CHART B.

One-and-one-half-finger Abdomen.

$1/16$ pelvic bone...........132-egg type
$1/8$ pelvic bone...........120-egg type
$3/16$ pelvic bone...........109-egg type
$1/4$ pelvic bone.......... 98-egg type
$5/16$ pelvic bone.......... 87-egg type
$3/8$ pelvic bone.......... 75-egg type
$7/16$ pelvic bone.......... 64-egg type
$1/2$ pelvic bone.......... 53-egg type
$9/16$ pelvic bone.......... 42-egg type
$5/8$ pelvic bone.......... 30-egg type

$^{11}/_{16}$ pelvic bone.......... 19-egg type
$^{3}/_{4}$ pelvic bone.......... 8-egg type
$^{13}/_{16}$ pelvic bone.......... 0-egg type
$^{7}/_{8}$ pelvic bone.......... 0-egg type

MALE BIRD—CHART C.

Two-finger Abdomen.

$^{1}/_{16}$ pelvic bone..........180-egg type
$^{1}/_{8}$ pelvic bone..........166-egg type
$^{3}/_{16}$ pelvic bone..........152-egg type
$^{1}/_{4}$ pelvic bone..........138-egg type
$^{5}/_{16}$ pelvic bone..........124-egg type
$^{3}/_{8}$ pelvic bone..........110-egg type
$^{7}/_{16}$ pelvic bone.......... 96-egg type
$^{1}/_{2}$ pelvic bone.......... 82-egg type
$^{9}/_{16}$ pelvic bone.......... 68-egg type
$^{5}/_{8}$ pelvic bone.......... 54-egg type
$^{11}/_{16}$ pelvic bone.......... 40-egg type
$^{3}/_{4}$ pelvic bone.......... 26-egg type
$^{13}/_{16}$ pelvic bone.......... 12-egg type
$^{7}/_{8}$ pelvic bone.......... 0-egg type

MALE BIRD—CHART D.

Two-and-one-half-finger Abdomen.

$^{1}/_{16}$ pelvic bone..........200-egg type
$^{1}/_{8}$ pelvic bone..........185-egg type
$^{3}/_{16}$ pelvic bone..........171-egg type
$^{1}/_{4}$ pelvic bone..........156-egg type
$^{5}/_{16}$ pelvic bone..........142-egg type
$^{3}/_{8}$ pelvic bone..........127-egg type
$^{7}/_{16}$ pelvic bone..........113-egg type
$^{1}/_{2}$ pelvic bone.......... 98-egg type
$^{9}/_{16}$ pelvic bone.......... 84-egg type
$^{5}/_{8}$ pelvic bone.......... 69-egg type
$^{11}/_{16}$ pelvic bone.......... 55-egg type
$^{3}/_{4}$ pelvic bone.......... 40-egg type
$^{13}/_{16}$ pelvic bone.......... 26-egg type
$^{7}/_{8}$ pelvic bone.......... 11-egg type
$^{15}/_{16}$ pelvic bone.......... 0-egg type

MALE BIRD—CHART E.
Three-finger Abdomen.

1/16 pelvic bone..........235-egg type
1/8 pelvic bone..........220-egg type
3/16 pelvic bone..........205-egg type
1/4 pelvic bone..........190-egg type
5/16 pelvic bone..........175-egg type
3/8 pelvic bone..........160-egg type
7/16 pelvic bone..........145-egg type
1/2 pelvic bone..........130-egg type
9/16 pelvic bone..........115-egg type
5/8 pelvic bone..........100-egg type
11/16 pelvic bone.......... 85-egg type
3/4 pelvic bone.......... 70-egg type
13/16 pelvic bone.......... 55-egg type
7/8 pelvic bone.......... 40-egg type
15/16 pelvic bone.......... 25-egg type
1-in. pelvic bone.......... 10-egg type
17/16 pelvic bone.......... 0-egg type

MALE BIRD—CHART F.
Three-and-one-half-finger Abdomen.

1/16 pelvic bone..........257-egg type
1/8 pelvic bone..........242-egg type
3/16 pelvic bone..........227-egg type
1/4 pelvic bone..........212-egg type
5/16 pelvic bone..........197-egg type
3/8 pelvic bone..........182-egg type
7/16 pelvic bone..........167-egg type
1/2 pelvic bone..........152-egg type
9/16 pelvic bone..........137-egg type
5/8 pelvic bone..........122-egg type
11/16 pelvic bone..........107-egg type
3/4 pelvic bone.......... 92-egg type
13/16 pelvic bone.......... 77-egg type
7/8 pelvic bone.......... 62-egg type
15/16 pelvic bone.......... 47-egg type
1-in. pelvic bone.......... 32-egg type
17/16 pelvic bone.......... 17-egg type
1 1/8 pelvic bone.......... 0-egg type

MALE BIRD—CHART G.

Four-finger Abdomen.

$1/16$ pelvic bone	280-egg type
$1/8$ pelvic bone	265-egg type
$3/16$ pelvic bone	250-egg type
$1/4$ pelvic bone	235-egg type
$5/16$ pelvic bone	220-egg type
$3/8$ pelvic bone	205-egg type
$7/16$ pelvic bone	190-egg type
$1/2$ pelvic bone	175-egg type
$9/16$ pelvic bone	160-egg type
$5/8$ pelvic bone	145-egg type
$11/16$ pelvic bone	130-egg type
$3/4$ pelvic bone	115-egg type
$13/16$ pelvic bone	100-egg type
$7/8$ pelvic bone	85-egg type
$15/16$ pelvic bone	70-egg type
1-in. pelvic bone	55-egg type
$17/16$ pelvic bone	40-egg type
$1 1/8$ pelvic bone	25-egg type
$1 3/16$ pelvic bone	10-egg type
$1 1/4$ pelvic bone	0-egg type

We consider the male bird of so much importance that we have made seven charts for his classification as to egg and meat types. See Charts A, B, C, D, E, F, and G. While Chart A may not be needed and Chart B used very seldom, we thought it best to include them. All old poultrymen and stock-raisers know that so many considerations enter into the breeding and raising of live stock of all kinds that it is impossible to lay down hard-and-fast rules that can be depended upon beforehand to bring definite results in all parallel cases. This is written as a caution to beginners, especially to those whose experience has been at the desk or behind the counter.

Fig. 46 shows a cock bird four fingers abdomen and Fig. 47 shows the same bird $1/8$-inch pelvic bone, making him a 265-egg type bird.

The reader will see by Figs. 46 and 47 that we use the same methods to determine the egg-value of a male bird as we use for the hen, except that we do not think it advisable to take the matter of condition into consideration, or rather it is better not to lay down rules in the matter, as it is very hard to keep

Fig. 46—Showing four-finger depth of abdomen of 265-egg cock bird.

Fig. 47—Showing ⅛-inch pelvic bone of 265-egg cock bird.

THE CALL OF THE HEN. 99

Fig. 48—Showing 1/16-inch pelvic bone of 280-egg type hen.

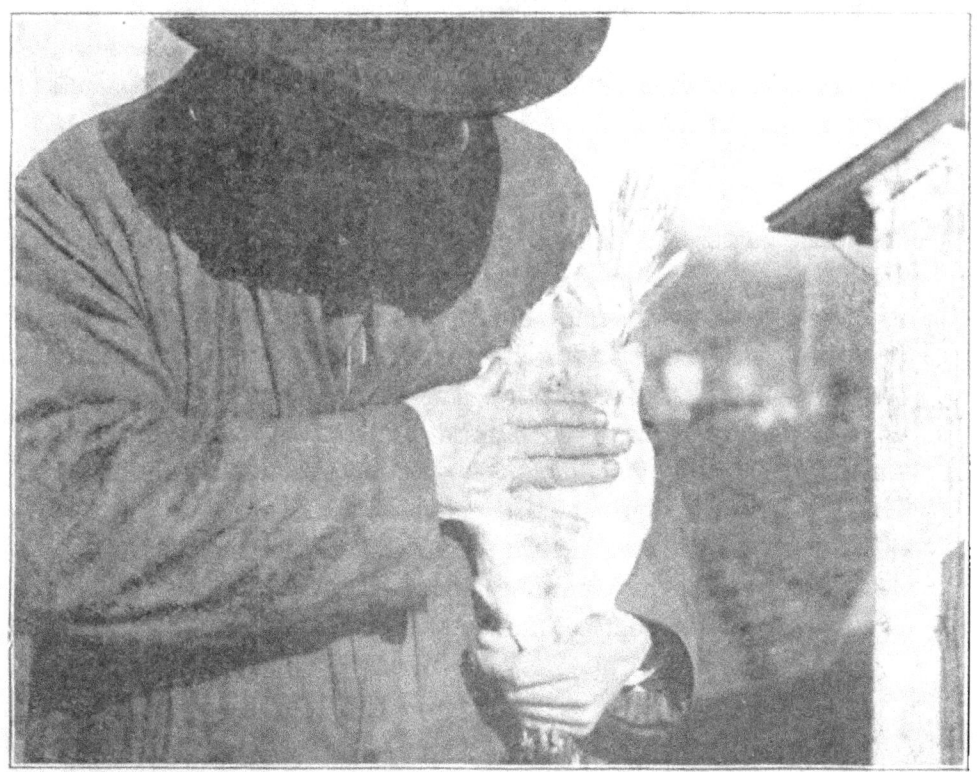

Fig. 49—Showing six-finger depth of abdomen of 280-egg type hen.

Fig. 50—280-egg type hen and 265-egg type cock bird. Tail of cock is somewhat cramped for want of room.

the egg-type birds in good condition; but I try to keep my cock birds in good flesh and not over one finger out of condition at any time. There are times before the male birds are a year old and while their bones are soft that their abdomens will contract and expand, it depending on whether they are stinted in their feed or whether they are fed liberally. Egg-type cockerels selected for breeders should have the best care and food (see chapter on Selecting Cockerels for Breeding). In examining the male birds for prepotency, the reader should select them with the greatest care. I cannot impress this on the reader too strongly. They should be as good or better if possible than No. 1, Fig. 35, and do not forget that the thumb-nail on the left hand and the nail on the forefinger of the right hand (reverse the order if left-handed) must be somewhat longer than the flesh, if you expect to take correct measurements.

CHAPTER XII.

Selecting the Cockerels at Broiler Age.

I have tried to impress on the reader the importance of the careful selection of the male birds, and perhaps he is fully alive to the value of doing so. He starts out at the first opportunity and visits all the poultry plants far and near, with the determination to purchase the best male bird he can find. Before starting out, he decides he will have nothing less than 200-egg types. Imagine his disappointment when, after handling perhaps fifty or more, he can find nothing that will come any way near the 200-egg type; while if he examines the same number of hens, he will very likely find at least one or perhaps more that will come somewhat near what he is looking for. Then he will say that there is no such bird as the chart describes as a 200-egg type cock bird. I wish to say here that I think I have at least fifty male birds at the present writing that will scale from 200 up, according to the charts. I have over a dozen that will scale from 250 to 265, and these have all been developed within six years from hens with three-finger abdomens and ¼-inch pelvic bones, mated to cockerels with 1½-inch finger abdomens and 1/16-inch pelvic bones.

The first season in California we raised about 300 cockerels up to three months of age, which is within the broiler age for this section. We arranged our house and catching-coop as in Figs. 1 and 2, and we went through the same movements that we do when testing the hens, except that we do not have to use all the tests on each one of the cockerels that we use on the hens. We hold the cockerel as in Figs. 5 and 6 and lay our hand on his abdomen as in Fig. 7. As soon as we lay our hand on his abdomen we can feel instantly whether his pelvic bones are straight, like Fig. 34, or crooked, like Fig. 33. If his pelvic bones are like Fig. 33, we have no use for him as a breeder and put him in the shipping-crate for market; if his pelvic bones are straight, like Fig. 34, we measure the depth of his abdomen; if it is less than two fingers, we put him in the shipping-crate; if two fingers or over, we examine him for prepotency; and if the projection on the back of his head, as in No. 1, Fig. 35, is less than ⅛ of an inch behind a line drawn at right angles from the back of the ear (see Figs. 41, 42, and

43) we put him in the shipping-crate, no matter how good he is in other points. We take no chances with him, because, **if we have made no mistake in measuring his head lines,** abdomen, and pelvic bones, it will be a waste of time to breed from him; but if his head measures up good, we keep him as a prospective breeder. We say "as a prospective breeder," as it is very evident it will not pay to raise all the cockerels to maturity.

Here in Petaluma, where there are over 600,000 cockerels raised to broiler age in a season, it would be impossible to raise them all and test their breeding qualities, neither is it necessary. If a person has a **delicate touch, the comparative value of chicks for prepotency** can be judged as well when they are **three days old** as at any time later. Then again, we are obliged to keep our chicks until we can distinguish the males from the females, and as a rule we will lose nothing if we keep them until they are at least ten weeks old, when, if they have had the right care and feed, they will be old enough to test. If their pelvic bones are thick at this age, it indicates they are more or less of the meat type; if their pelvic bones are crooked, it indicates that they never will be straight; and if they lack prepotency, it indicates that they will always lack it, for they come out of the shell with this organ relatively large or small, just as a baby is born with a nose on its face.

I want to impress on the reader the importance of using the utmost care in measuring the head for prepotency, as it is very easy for a person **to think** he has measured the head right when he has not done so; especially if he has large self-esteem, he then thinks everything he does must be right; it would be impossible for **him** to do anything otherwise than the right way. In my classes I have found workers in the machinists' trade made the most correct measurements, especially if they had the faculty of human nature large, while I have found that professional men who had human nature small make the poorest measurements; this was owing to prejudice, and not to the absence of the combination of the necessary mental faculties. I suppose there will always be found those who will discredit the most obvious fact, if it puts them at a disadvantage from a mental, moral, or financial point of view; but in this case it would be cutting off your nose to spite your face to be careless in any of these tests.

I have never yet, in my investigations of hundreds of poultry plants, found a degenerate lot of poultry but that they were **small in prepotency.** But to return to the cockerels: As we said on page 101, we raised 300 cockerels the first year I was in California. After testing them at three months old, as described, I found eighteen that I considered worth keeping to the age of nine months, when I would give them the final test. When they were eight months old I tested them again, and while I found that they all had good depth of abdomen and good prepotency, six of them had crooked pelvic bones. The pelvic bones on twelve of the cockerels had continued to grow straight, while the pelvic bones on six of them had grown crooked and were coming together at the points, like the horns on a Jersey cow. I had to discard these six breeders and send them to market.

The reader will see that, **out of 300 cockerels, I had only 12 that were capable of improving my flock.** Last year (1912), out of about 1,200, I had only 200 that I considered good enough to keep for breeders; and while all my birds have been more or less squirrel-tailed, one of last year's 200 is a very well-formed, low-tailed bird, but he lacks the pure-white ear-lobes. He scores 250-egg type, and I have refused $50.00 for him. I am going to see if I can breed a low-tailed type of Leghorn in quantities that will conform to the present American standard, and average about 200 eggs per year in large flocks. The reader will understand that the parents of these cockerels were selected with the greatest care as to capacity, type, and prepotency. **Type and prepotency** are more or less hereditary traits or features, distinguishable in the subjects, if we have the knowledge necessary to discern them. **But the individual inherent or innate potentiality** of any one or each bird cannot be increased or diminished by the breeder; that is to say, feed and environment will not materially change the **impotent** bird into a **potent** bird, neither will it change the typical **meat-type** into the **egg-type** bird.

"But," I hear some sarcastic reader say, "we certainly can diminish or increase their prepotency by alternately starving and feeding them well." That is begging the question. You could affect their fecundity very readily; but what the writer wishes to impress on the reader is, that while **type** and **prepotency** are fixed before birth, and also the **ability** to govern capacity, and while **type** and **prepotency** can be procured **only**

by selection, capacity can be governed more or less by environment—in other words, feed, care, the right kind of houses, ground, etc. We will say, for instance, the reader has a pen of egg-type birds, both male and female, with large prepotency and capacity, and suppose they were all 200-egg birds. There would be no difficulty in raising chickens from them with the same degree of type and prepotency; but if he should stint them in feed of the **proper kind and quantity** while growing, they would lose in capacity each generation. I develop the capacity of both pullets and cockerels from the time they are three days old to the fullest extent by the most liberal feeding, care, and surrounding conditions.

In concluding this chapter, I would say that the bird with the desired characteristics is more or less of a sport, and the value of the **"Hogan Test"** lies in the fact that **with this knowledge you can discover the sport** and perpetuate it through intelligent breeding. Again, I want to say here that my **best** cockerels measure four fingers abdomen at three months old. All my stock is developed as much as possible at this age, and I try to prevent the cockerels from shrinking. But the pullets will develop until some of them are six fingers abdomen.

The enclosed article from the Petaluma *Weekly Poultry Journal* emphasizes what we have said in regard to the feeding and care of young stock. These cockerels were not crammed or penned up and fed, but were taken off free range and sent directly to market. I wish to remind the reader here that in examining the cockerels for prepotency he may be proficient enough in the matter to examine them by holding them between his knees and not be obliged to put each one in a sack. The article follows:

"Walter Hogan Can Raise Chickens.

"Walter Hogan backs up his system of selecting the good layers from among the poor ones, but he has never made much fuss about his ability as a poultry-raiser. For that reason some people have absorbed the idea that he is more of a theorist than a practical man. But he now has a flock of his own, and evidently he is making good, for he is getting results that will convince any one from Missouri or anywhere else who must be "shown" before believing. For instance, last week there **was a spell of most discouraging depression in the prices which dealers were willing to pay for young poultry.** There were large arrivals of Eastern poultry in San Francisco besides heavy receipts of California, and nobody wanted any more. Just the same, Mr. Hogan received $4.00 a dozen for sixteen dozen cockerels just three months old, when the same dealer was paying but $1.50 for birds of the same age. Now, what do you think of that? And Mr. Hogan says these cockerels were not descendants of the beef type of hens, but were hatched from eggs laid by hens selected as the egg type. They were not especially fed or in any way prepared for market. They cost 22 cents each for feed, and thus the profit on the bunch was $21.76.

"In speaking of this matter, Mr. Hogan made the point that if all poultrymen would pay especial attention to producing fine broilers for market—that is, in preparing the broilers that they are obliged to produce in order to have a corresponding number of pullets—they would benefit themselves greatly. Not only would they get a better price for the birds, but they would greatly increase the demand, as many people who now care nothing for the common dry-meated birds would become pleased consumers of the improved broilers. The *Poultry Journal* man knows by personal experience that the broilers turned out by Mr. Hogan are simply delicious when properly cooked, and far ahead of the ordinary article."

CHAPTER XIII.

Selecting the Setting Hen.

"How can I select the best hen for the purpose when I want to hatch chickens with hens?"

The writer is asked the above question very often. It is a serious matter with a poultryman when he has a small number of choice eggs he wishes to hatch and gives them to a hen that is apparently setting well only to have her spoil most of them. He very naturally lays the cause to mites or lice, or both. While it is true that the nests and surroundings must be kept free from mites and the hens kept clean from hen lice, the trouble is not all here by a good deal. Sometimes a great deal of the fault lies in the hen. Some are born layers, some are born mothers, and some are born too lazy to get off of the nest at the call of Nature. The hen born a typical egg type is of no use as a setter, neither is the hen that is born a typical meat type, she is too lazy to care for her chicks, even if she is fortunate enough to hatch any and not kill them all by standing on them. She is too stupid any way, and the typical egg-type hen is too nervous and has no time to attend to them. She thinks of nothing but manufacturing eggs. So we will have to look for a hen between the above types, which we have in the **dual-purpose type,** with the following characteristics:

First, she must have **prepotency large;** that gives her the mother instinct; next, she should be in normal condition, as **indicated by her breast-bone;** that is self-evident, for a hen out of condition lacks more or less of the animal magnetism that is an aid to successful incubation. I need not mention good health, as indicated by good red comb and wattles, as everyone knows that. The hen should be four fingers abdomen, since anything heavier is more or less liable to break the eggs and anything less than that would not be large enough to cover sufficient eggs. If the hen is a three-finger abdomen hen, her pelvic bones should be about $7/16$ or $1/2$ of an inch thick; if she is a four-finger abdomen hen, her pelvic bones should be about $1/2$ or $9/16$ of an inch thick. If you can find hens such as described here, you will have hens with the mother instinct. They will not be too lazy to take proper care of themselves and their chicks, nor will they want to lay so soon as to neglect their chickens. The nearer you can get to procuring the above type of hens the better success you will have raising chicks with them.

CHAPTER XIV.

Selecting the Stock for Raising Broilers.

A great many of my friends have requested me to write a chapter on how to raise broilers, but as there are so many excellent books on the market that describe the process of the feeding, caring for, and raising of broilers a great deal better than I could do it, I will confine myself to the selection of the breeding stock only. The writer has raised Light Brahmas and White Plymouth Rocks for years, and has experimented with them to get the greatest amount of meat from the smallest amount of feed; to get the greatest weight of meat at three months old in the White Rocks and the greatest weight of meat in the Light Brahmas at maturity. In the process I have run up against two distinct propositions; one was a success from a commercial point of view, and the other, while not a financial success, was a success from an epicurean point of view. I will describe the financial proposition first:

We will select a pen of hens from our favorite breed, or from Wyandottes, Orpingtons, Plymouth Rocks, or Rhode Island Reds. The hens **must have large prepotency;** they must be six or seven fingers abdomen and their pelvic bones should be ⅝ of an inch thick, in good condition. Now you have hens that should lay twelve dozen eggs their first laying year, and they are a paying proposition. Do not breed from them the first year, but wait until they are over one year old; then mate them with a mature cockerel or young cock with **large prepotency,** with abdomen four fingers deep **or more** and pelvic bones from 1 inch to 1¼ inches thick. You should feed the pen for eggs, and keep them as healthy as possible. If they are fed right, you will get lots of eggs and good, healthy chicks, capable of putting on flesh rapidly and fattening very easily. As a **paying proposition** for market broilers, I have never found any combination that would equal it.

But for my private use, without regard to profit, I would take the same combination as the above, **except** that the **pelvic bones** of the hens would be **1 inch thick,** instead of about ⅝; this would give a broiler that would put on flesh much faster, consequently it would be more tender. I have raised broilers the flesh of which would almost melt in your mouth. I have a few secrets in the raising of them which I have never divulged, but may do so in a few years.

Fig. 51—The dry-mash hopper we use, closed.

Fig. 52—The dry-mash hopper we use, open.

CHAPTER XV.

Using the Hogan Test in Judging Poultry at the Poultry Shows.

From the *Live Stock Tribune*, Los Angeles, California.

(Now *Pacific Poultrycraft*.)

"Inglewood Poultry Show.

"A poultry show will be held in the Inglewood Poultry Colony on March 13th and 14th. This show will be the first of its kind ever given to the United States. All poultry shows that have been held in this country heretofore have awarded prizes according to the color, markings, and shape of the fowls only. The show at Inglewood will be unusual in that prizes will be awarded irrespective of the color, variety, shape, size, or age of the fowls in competition.

"Birds in competition will be judged as to their egg-laying capacities and reproductive abilities only. The judging will be done by the system discovered and perfected by Walter Hogan and now used in practical poultry-raising by the members of the Inglewood Poultry Colony.

"First, second, third, fourth, and fifth prizes will be awarded to the best males and females entered from Inglewood; first prize being $5.00 cash, second prize being $3.00 cash; all winners receiving ribbons. In addition to the foregoing, there will be the Jaffa Grand Prize of $25.00 in gold, which will be awarded to the hen in the show which shows the greatest capacity as a layer, combined with the ability to reproduce her kind.

"Entries for the regular prizes will be limited to fowls from Inglewood, but competition for the Jaffa Grand Prize will be open to all comers. Entries from poultry-raisers outside of Inglewood will be limited to two birds each. No entry fee will be charged, but all birds entered will be sent at the owner's risk, as is usual at all shows.

"The birds entered will be cared for and reshipped to the owners by White Wyandotte Farm, under whose auspices the show will be given and to whom all entries should be sent. No entries will be received after 10 o'clock A. M. on March 12th.

"This show will be unique in that it will present the commercial side of the poultry industry to the exclusion of fancy breeding. Every step in the poultry business from the hatching of the chick to the preparation of the mature fowl for market, and the packing of the eggs for table use will be illustrated by actual demonstrations on the famous White Wyandotte Farm, where the exhibition will be given. Incubators will hatch not less than 2,000 chicks during the show, and chickens in every stage of development, from one day old to ten weeks old, will be shown as raised in the best brooders with the best care.

"There will be demonstrations on both days of the show of killing, picking, and preparing fowls for market, as well as of packing fancy eggs. The best and latest in poultry supplies, fittings, and equipment will be shown as actually used by the capable, successful men who are in the business for revenue only.

"No admission fee will be charged, the show being given for the purpose of exploiting and demonstrating the poultry business as it is being developed in Southern California.

"The Jaffa Grand Prize is given and named in honor of Professor Jaffa, of the University of California, who was the first man in public life in this State to test and verify the excellence of the system discovered by Mr. Hogan.

"Transportation from Los Angeles to Inglewood will be free, and it is understood that the Board of Trade of Inglewood will make arrangements to take those who visit the show around the city of Inglewood in automobiles.

"Those who visit the Inglewood Poultry Show will see an exhibition that will be more interesting by far than any show that has preceded it in California or in any other State, because one will have an opportunity to see, not the pedigree, but the money in the chicken and a practical way to get that money out."

In judging the poultry show at Inglewood the management made the rule that all birds were to be judged according to the condition they were in at the time they were judged, and while this rule may be all right in judging the fancy bird and the beef-type bird, it will never do for the egg-type bird, as the reader will see when I relate an incident that occurred during the show in Inglewood, which was held in March. A gentleman had entered a White Leghorn hen that he had trap-nested a year up to the previous November, and had her record with him. The hen scored (as near as I can remember) two fingers abdomen, two fingers out of condition, and $^3/_{16}$-inch pelvic bone, and according to the rules of the show I was obliged to give her credit for 78 eggs her first laying year, when, according to his trap-nested record, she had laid 180 eggs. He said she had been sick and had just commenced to improve shortly before he sent her to the show, and he wanted to prove whether or not I could tell how many eggs she **had** laid her first laying year. I told him I could not tell how many eggs she **had laid,** but I could tell how many **she could have laid** if she had been fed and cared for right, barring accidents and sickness; that her capacity was 190 eggs her first laying year. He then showed me her record, which was 180 eggs.

In the autumn of 1911 George D. Holden, ex-president of the American Poultry Association, judged the fancy and the writer judged the utility birds at the Pajaro Valley Poultry Show, held at Watsonville, Santa Cruz County, California. In judging that show full credit was given each bird, **both male and female,** as to what they were capable of doing, whether in meat or eggs, and for prepotency, without any regard as to how their owners cared for them—or, in other words, without regard to their condition. And the owners of the birds who were interested in knowing were instructed how to rectify any deficiency there may have been in the birds. It seems to me this is the best way to encourage and develop the poultry industry. I am sure the American Poultry Association could formulate a code of rules that would **greatly** aid in judging utility poultry and thereby add greatly to the interest of our poultry shows; in fact, I am advised that such a proposition is being considered at the time I am writing this (July 25, 1913).

CHAPTER XVI.

Stamina in Poultry.

When I came to California and told the poultry-raisers that I was going to take their birds and in the course of time breed a flock of 200-egg hens from them, they declared it **could not be done.** They said if it was possible to breed up a large flock of 200-egg hens, their progeny would be so weak I could never raise them, and that their eggs would be so misshapen and thin-shelled they would not be marketable. I replied that perhaps they were right, but I saw no reason why I could not do so here, as I had bred up one lot in the Eastern States and another lot in Minnesota. Both lots were Leghorns, and I thought it would be easier to develop Leghorns in California than in Minnesota, and I have now demonstrated in California that the following can be done:

1. The 200-egg hen is a fact and not a theory;
2. That she can be **bred and fed** to lay as perfect an egg as any other class of hens;
3. That her eggs are as fertile and will hatch as strong chicks as the hen that does not pay for her feed.

The breeder need not take my word for the above statements. The frontispiece shows five of this type of birds that the writer bred and raised in California. These birds laid the greatest weight of eggs (131 pens of five birds to each pen competing, including three pens of Indian Runner ducks) in the National Egg-laying Contest at the State Poultry Experiment Station, Mountain Grove, Missouri, U. S. A., for the twelve months ending November 1, 1912. These five hens laid 131 pounds of eggs, which, reduced to No. 1 eggs as rated in Petaluma, would be $229^3/_5$ eggs for each hen. The eggs these five hens laid **while moulting** were put on exhibition in the Chamber of Commerce in Petaluma and were pronounced by good judges to be as fine a lot of eggs as they ever saw, and that is saying a great deal, as there are more eggs produced within a radius of ten miles from Petaluma than in any other like part of the world. We have hundreds of letters from our customers testifying to the value of this stock, a few extracts from which we will introduce here to prove to the reader that because a flock of hens are great layers it does not follow that they are of low vitality.

EXTRACTS FROM LETTERS.

PORTLAND, ORE., June 23, 1912.

Received eggs. None broken. Very nice. **Fifteen infertile** out of 150. C. F. PERKINS.

LIHUE, HAWAII, June 11, 1913.

Eggs arrived O. K. None damaged. Have fourteen chicks four weeks old doing fine. Am well pleased.

E. H. BROADBENT.

(These eggs were shipped over 2,200 miles by rail and steamer to reach their destination.)

WATSONVILLE, CALIF., April 5, 1912.

Eggs received. Finest we ever had. Got forty-nine strong chicks from sixty-four eggs. ORA L. HILL.

VANCOUVER, B. C., May 13, 1912.

The 100 eggs received. Express and customs ran price to $14.00. Am very well satisfied. Hatched 70 per cent beautiful chicks; doing well. G. W. McLELLAND.

QUINCY, WASH., April 14, 1912.

Chicks received; not a dead one in the bunch, which speaks well for the vitality of your stock.

H. L. JOHNSON, Treasurer
and Manager Quincy Lumber Company.

VICTORIA, B. C., Sub. P. O. No. 1,

April 19, 1912.

Received the 100 chicks; four dead. Think that is very good, coming that journey. JAMES D. WEST.

SALEM, ORE., April 19, 1913.

Received baby chicks; they are just lovely; not one dead, which we think is great. They came in fine shape.

MR. and MRS. HAYRE.

SEATTLE, WASH., August 25, 1912.

Received the 1,040 chicks about ten weeks ago; there were five dead in the boxes. Have lost about 75 of them, all told.

S. K. SUTTLE.

TUCSON, ARIZ., February 17, 1913.

Received chicks in good condition; 1 dead, 623 alive and kicking. L. E. SMITH.

RENO, NEV., March 11, 1913.

Chicks came through fine; 1 dead in 700, which speaks well for their vitality. They surely are a spry bunch.

A. L. RICE.

RENO, NEV., July 22, 1913.

Chicks are fine; they are the largest and best-looking ever seen in Nevada. They are just 4 months and 12 days old. One of them laid yesterday. Every poultryman that sees them remarks it's too bad I haven't a thousand.

A. L. RICE.

The preceeding extracts are taken from a few of the many unsolicited letters I have received from my customers during the last two years that I have been selling hatching eggs and day-old chicks. I have repeatedly shipping hatching eggs to the Hawaiian Islands and as far east as Minnesota, and day-old chicks where they would be over seventy-two hours on the road. Last summer I turned down over $6,000 worth of orders that I could not fill at $10.00 per 100 for eggs and $15.00 per 100 for day-old chicks. I am aware I will have a hard time convincing some of my readers that what I claim for the 200-egg hen is true, but it seems to me any progressive poultryman would be satisfied with the proof I offer him. I will admit that the eggs and chicks from the 200-egg type hens as now bred are not all we would desire, but that is owing to lack of proper knowledge of breeding. As I have said before, by using the "Hogan Test" the reader can breed as fine or as coarse as his conditions require; and by selecting only those birds with large prepotency he will be assured of success.

CHAPTER XVII.

"At Sea Over Mating"—What Shall It Be, the Trap-Nest, Mendelism, or the Hogan Test?

(From *The North American*, Philadelphia, Pa., November 24, 1912.)

"At Sea over Mating.

"America has some good layers, unheard of and unknown, 'tis true, but we are evidently all at sea in the matter of mating for egg-production.

"Can it be possible that Mendel's law obtains in egg-production just as it does in feathers and form? Do we eliminate, according to Mendel, the factor governing certain things in egg-production, just as we do in the attempt to control coloring in birds, fowls, animals, and flowers? If a a son of a heavy-laying female is mated to a non-layer and this son does not carry the excess of laying proclivity, do we get poor layers or good layers? If a 100 per cent producing hen (200 eggs or more) is mated to the son of a 100 per cent producing female, it does not follow, if Mendel's law applies, that the mate to the second 100 per cent female inherited egg-laying proclivities; therefore, why should the offspring of the second mating be prolific egg-producers? And how far back must we go to get the excess of female inclination to reproduction?

"Predominance of inclination exists somewhere in some tangible form, but we do not seem to be able to find it under our present system. That we will is conclusive, but we must do so quickly, in order to offset the growing increase of foodstuffs."

The trap-nest identifies and gives you the number of eggs a hen lays and is absolutely necessary if we wish to line-breed or raise pedigreed stock. The writer has studied Mendelism since the spring of 1910, as he has numerous other scientific works, in the endeavor to find something that would be of aid in getting out this work. I must confess that the title, "The Call of the Hen," was suggested while on a visit with Comrade Jack London, and that is all I have been able to find that has aided me in this case. Mendelism may be found an aid along the line of feathers, but I doubt if there is anything in it that will aid the poultryman in the selection of breeders for type, stamina, and the production of eggs or meat. It may be that, having eyes, I fail to see it. Even if there should be anything of value in Mendelism, it would take two or more years to get it out, while "The Hogan Test" indicates the value of a bird in a few minutes, at most. It looks to me as if the poultrymen will have to look at the trap-nest and "The Hogan Test" to develop and maintain the high-scoring meat- and egg-producing hen. The best pullets can be selected at maturity by "The Hogan Test" and then trap-nested when the poultryman is breeding pedigreed stock; while the culled pullets, lacking in prepotency and other points, can be kept as market-egg producers. In this way it will be neecessary to trap-nest only the cream of the flock, and thereby save an immense amount of labor. The cockerels can also be selected at three months of age and the most promising saved from slaughter. By this method poultry-breeding will be reduced to a science and become a pleasure, where now it is a brain-racking proposition.

CHAPTER XVIII.

"How Can I Tell a Laying Hen?"

I am asked this question very often, and in reply would say that from a **scientific** point of view it is impossible to tell the laying hen except with the X-ray. When I say this I do not mean that you cannot tell in the vast majority of cases, but there are occasionally hens whose formation is such that no known method will tell you whether she is a laying hen or not. I give in the last chapter my original "System" and the later supplement, which caused a great many questions to be asked, which I trust have been satisfactorily answered in this book.

I was at a place in San Francisco lately where this subject was brought up. There was a small party present, all of whom had my "System." One of the party worked in a large meat-market, where they bought and dressed live poultry. He said that occasionally he dressed a hen that showed no indications of being a laying hen, but upon being opened an egg would be found in her. I told him the hens that he had described were those that laid a very few eggs and laid them only in the **spring.** Their pelvic bones expanded only while the hen was being delivered of the egg. This hen has **practically** but one **egg** under process of development at a time, consequently her abdomen does not have to expand to make room for only one egg. Whereas the hen that lays 150 eggs per year has a number of eggs developing at the same time, and her abdomen expands in proportion to her needs. The 200-egg hen has a still larger number of eggs developing and she requires more room for them, hence her abdomen expands in proportion. The 250-egg hen has a still larger number of eggs of all sizes developing, and her abdomen expands still wider than the 200-egg hen. When the hen's abdomen expands, her pelvic bones, being literally a part of and continuation of her abdomen, must expand and contract with it. When she is through laying for the season her abdomen contracts, and the pelvic bones must come closer together, which they do, although there are exceptions to this rule. We will take the 145-egg hen, for example, of the sanguine temperament. She will be four fingers abdomen and ⅜-inch pelvic bone, when in normal condition, with pelvic bones of good shape. We draw our hand along her breast-bone (keel) from front to rear, and find her abdomen does not drop down the least bit below the rear of the breast-bone. This hen we might call a "normal hen." Her pelvic bones will, in all probability, expand and contract in conformance with her condition of laying. If she is in the flush of laying, her pelvic bones may be about 1¾ inches apart; later in the season, when she is not laying so frequently, her pelvic bones may close to about 1½ inches; and when she stops laying for the season her pelvic bones may close to about 1¼ inches. This will very likely be repeated each year.

Now we will select a hen of the 250-egg type. We draw our hand along her keel, as with the last hen; we find she is all right, closely built and firm. We drop her and take another 250-egg type hen. The performance of drawing the hand along the keel is for the purpose of picking out the future breeders that may later bag down, indicating weak ovaries. In this connection I wish to say that in selecting breeders I found that the best way to eliminate the hens that would begin to bag down behind was to follow directions as given below. Of late years I have not had this trouble to contend with. It is always the heavy layer that breaks down, which indicates weak ovaries, and we do not want to breed from such.

In drawing our hand along her keel (breast-bone) we find a slight bagging down in the rear. The abdomen seems to drop below the rear of the breast-bone slightly. We will say this is a pullet, perhaps six or eight months old. She is well developed, and you can call her one of your best hens. You are proud of her, and have decided to set every egg she lays. Do not use her as a breeder. This pullet should be put in a yard with others of her formation after she is sixteen months old and trap-nested. She may stop laying any time and never lay another egg, or she may continue to lay another year or so; in any case, she has been such a continuous layer that her frame has become set to that form, and her pelvic bones, as it were, set and will contract very little; they will indicate that she is laying, when in fact she may not have laid for years. I have kept such hens until they were 6 years old, and some of them have never laid an egg after they were **about** 16 months, still others after they were 2 years old. This is where a trap-nest will save you money. When you select your hens by Charts 44 and 45 at 16, 28, and 40 months of age, the ones that bag down the least bit should be put in a yard by themselves and trap-nested to discover the ones whose ovaries have broken down and will lay no more. This is not difficult to discover, as the hen that is over the 205-egg type lays more or less at all times during the first two years of her life, if not stimulated to over-production her first year. "A little learning is a dangerous thing," is an old saying applicable to this case. When a man says, "Don't kill that laying hen," he should furnish you with an X-ray outfit that will enable you to comply with his request.

The writer has used the pelvic-bone proposition for over forty years in selecting the laying hen, and has found the following to be a very good method in selecting the hen that is not laying:

The hen that scores 130 eggs her first laying year would measure about $7/8$ of an inch between her pelvic bones after she stops laying for the season. The hen that scores 150 eggs her first laying year would measure about 1 inch between her pelvic bones after she stops laying for the season. The hen that scores 200 eggs would measure about $1\frac{1}{4}$ inches between the pelvic bones after she stops laying for the season. The hen that scores 250 eggs would measure about $1\frac{1}{2}$ inches between the pelvic bones after she stops laying for the season. The 250-egg hen does stop more or less after her second and sometimes after her first season, if not cared for right; but if feed and environment are right, she may continue to lay more or less until 3 years old, when her frame may become set. When she is done laying her pelvic bones may remain 2 inches apart. As hens grow older their pelvic bones become thicker during the winter months when they are not laying. The thickness varies according to their type, the typical egg type changing little or none, while the more pronounced the meat type becomes the more the pelvic bones change, owing to the increase or decrease of flesh on the abdomen (flank) of the fowl as it takes on or loses flesh, as indicated by her breast-bone.

CHAPTER XIX.

FINAL REMARKS ON CONSTITUTIONAL VIGOR AND VITALITY.

As we have now reached the end of "The Call of the Hen," I wish to impress upon the reader's mind the importance of the five propositions that govern the Selection, Breeding, and Profitable Keeping of Poultry as follows: Capacity, Condition, Type, Prepotency, and Vitality or Constitutional Vigor.

No doubt you have a good working knowledge of the first four subjects, and you wonder why I have not written a chapter on Vitality. The reason is, that when I decided to write "The Call of the Hen," I told my wife that I would write nothing that even a blind man could not understand and practice. I have tried to do so, for to her patience, perseverance, and untiring zeal I owe much of the success I have had in getting out this book.

The writer can see only three ways of detecting vitality in a fowl; the most ancient is intuition, then observation, and lastly the trap-nest. A hen may be a typical 250-egg type hen, she may have the very best of care and food, and yet, for lack of vitality, may not be able to lay over 150 eggs per year. Let us take the steam engine for example. There are a great many types of engines besides the high- and low-pressure ones, as there are a great many types of hen and cock birds. The diameter of cylinder, length of stroke, and revolutions per minute give you the capacity of the engine, as the length and depth of abdomen in the fowl gives its capacity. The fuel fed into the fire-box generates the steam (vitality) to run the engine, as the food fed into the hen's abdomen generates her vitality.

The writer has owned steam engines where there was defective fire-box construction—scale in the boiler and tubes, loose rings in the piston head, cylinder worn out of true, and other defects that reduced the efficiency of the power system a great deal—or, in other words, lowered the vitality of the engine. In just the same way a weak digestive system in a 250-egg type hen will reduce her egg-yield. But do not think that you can make a 150-egg type hen in perfect condition lay 200 eggs by any of the feeding formulas now in vogue. If you try to force her, she will go to flesh and then break down with liver trouble.

If you lack the intuitive faculty and lack the time to carefully observe individual hens, I would advise you to select the hens by the chart you wish to breed from. When they are about a year old you can breed from them. Then, if you wish to breed from only those with the greatest vitality, trap-nest these hens for the next two or three years. The hens with the greatest vitality will be great layers and strong, vigorous birds, and save the time wasted in trap-nesting a lot of birds that you will eventually have to discard.

CHAPTER XX.

WALTER HOGAN'S SYSTEM.

This chapter contains "Walter Hogan's System," as written by M. F. Greeley, editor of the *Dakota Farmer*, to whom I gave the notes. This was published in 1904. At that time Mr. Greeley refused to put in anything about the skull theory. He said that I would make myself the laughing-stock of the world. I am merely putting this old work in this book in order that the reader may know the evolution of the discovery. The pelvic-bone method of selection was, of course, my first discovery; then later, the relation between depth of abdomen and thickness of pelvic bones; after that, the working out of the mathematical relation between egg-laying ability and those points before mentioned.

When I came to California I gave out merely the "Walter Hogan's System" which had been printed in Minnesota; later I published a "Supplement," which gave a general idea of the capacity and type proposition; still later I issued typewritten charts as they are found in this book. I could have done all of this many years ago, but my reasons for not doing it are explained elsewhere.

I do not desire any of my readers to make the mistake of considering what comes after this as having anything to do with "The Call of the Hen," except in a historical way.

WALTER HOGAN.

Petaluma, Calif., July 31, 1913.

WALTER HOGAN'S SYSTEM

WALTER HOGAN,
The Originator of the Walter Hogan System.

There are two ways of selection, described in this document.

When hens are in flush of laying, selection by the pelvic bones alone is the easier way; but when not in flush of laying, the pelvic bones together with the abdomen will be found the most ready way. (See Supplement, next page.)

Please bear in mind that the hen with thin pelvic bones and large, soft abdomen is the heavy egg-lying type.

The hen with thick pelvic bones and large, fleshy, fatty abdomen is the large beef type.

The hen with medium-thick pelvic bones and large, medium-fleshy and medium-fatty abdomen is the dual-purpose type, and can be made to lay fairly well or made to produce flesh, it being a matter of how she is fed.

The hen with small abdomen is of small account, either as an egg or as a meat proposition, as she lacks the abdominal capacity to digest and assimilate food enough to sustain the every-day wear of her system and at the same time to produce eggs or flesh in paying quantities.

Everything related here applies to the male bird as well, only in a lesser degree.

The remarks in regard to pullets refer to mature pullets, as Leghorn pullets are at five months old in the New England States.

My birds in Massachusetts were bred for eggs only for years, and their type became set and their pelvic bones contracted, when not laying, to average about 25 per cent; but I find that hens bred promiscuously contract about 50 per cent.

The points to be borne in mind in using this System are:

That selection by the pelvic bones alone is best made in the flush of laying.

That thin pelvic bones and soft abdomen indicate the egg type.

That thick pelvic bones and hard, fleshy, fatty abdomen indicate the beef type.

The size of the abdomen indicates the capacity of the bird, either as an egg or as a meat proposition, as the case may be—large abdomen, large capacity; small abdomen, small capacity.

The same rules apply to the cockerel, cock, male bird, or rooster, as he may be called.

In order to determine the capacity of a hen for egg-production by one selection, she should be in normal condition and not more than a few days broody.

The estimates in this document refer to hens about one year old. As a rule, they will lay less each year as they grow older—how much less depends on the vitality of the hen, other things considered.

Supplement to Walter Hogan's System.

If you will get a little 1-foot rule to check yourself up while getting used to measuring with the tips of your fingers, as in Fig. 4, you will have no trouble in applying its principles right. You can hold the bird feet up and head down between your knees while you are measuring; then hold as in Fig. 4 and learn to estimate the width right. Anything under 1 inch will not pay, all over 1½ inches will pay; from 1 to 1⅜ inches are doubtful; 2 inches is about the 200-egg type, 2⅜ inches about the 250-egg type, and 2¾ inches about the 280-egg type.

Hens measuring from 1 to 1⅜ inches should be put in a yard while being fed well and looked over once a week at night in the dark for about eight weeks, if you wish to make a careful test. Any that come up or down in measurement can be put in the good or bad yards, as the case may be. Hens will go up or down about 25 per cent in measurements as they are in flush of laying or not. The best time to examine hens is after dark while on roost, which should be about 18 inches from the floor. Place left hand on back of hen, lift up tail with thumb of right hand, and apply tips of fingers to pelvic bones. With a little practice you will be able to inspect thirty per minute. It is admitted by all physicians, professors, and students of physiology that I have talked with in regard to this matter that the abdominal capacity of a hen, together with a strong vital temperament, has everything to do with her value as a laying proposition. The pelvic bones (being a continuation of the body structure of the fowl and subject to very small changes in the formation of flesh) are, when comparatively straight and thin, an index to the width of the abdomen, and the best if not the only one we have, as they protrude from the body and may be easily measured. The depth of the abdomen can be taken by placing the palm of the hand crosswise below, between the pelvic bones and the rear of the breastbone. Sometimes it will be 1, 2, 3, 4, 5, or 6 fingers. (A finger means ¾ of an inch.) Also place fingers between pelvic bones and tail-bone. Sometimes it will take one, sometimes two fingers. In this way you can judge the size of the abdomen, which, with the pelvic development, will be a rule as to a hen's value as a layer, except in rare cases of misplaced or diseased organs. Sometimes a hen will have a large abdomen, but her pelvic bones will grow crooked and come almost together, like the horns of a Jersey cow, and she will lay better than the distance apart of her pelvic bones will indicate, but never will do as well as she should, and should not be bred from. She wastes too much nervous force in laying. The farther you get away from the crow formation the better your hens will be.

As a rule, fowls are almost twice as long coming to maturity in California as they are in the Eastern and Middle Western States. What the reason is I suspect, but do not know, but will find out in the next two years.

No document purporting to be a copy of "Water Hogan's System" is genuine without my signature as is set hereunder:

Wishing you the best of success, I am, sincerely yours,

The Walter Hogan System of Increasing Egg-Production by Selection and Breeding.

It has been estimated that to add one-half dozen eggs to the annual producing capacity of every hen in the United States would result in additional returns from our poultry sufficient to pay the national debt within less than a year. Allowing this to be true, we are prepared to show that the method of selection and breeding herein outlined is capable of paying off our great debt several times during a single year, without having to increase the number of hens kept a single bird or the cost of keeping them a single dollar.

The method—or "discovery," we might call it—has been tested by the writer in every conceivable way, regardless of expense, time, or trouble, and has been found absolutely faultless in every particular. It has been submitted to one Government Experiment Station (as will be shown later) with the same unerring results, and also to a number iof the foremost poultrymen of America, who fully and without exception corroborate all that is claimed.

This, you will agree with us, means a revolution in economical egg-production; it means, too, that no poultryman, however small his flock, can afford to go on in the old way a single year longer.

Every animal on the farm has a well-defined mission all its own, outside of the general one of producing meat. The great mission of the cow is to produce milk; of the sheep, wool; and the mission of the hen is evidently and pre-eminently egg-production. This being the case, her value varies or should vary largely with her ability to produce eggs. And still it is a well-known fact that, while every farm animal has been selected and bred for the best there was in it along its own peculiar line, and all prizes have been awarded accordingly, the hen has been bred largely and prizes awarded her almost wholly for feather and markings, the judges seldom or never deeming it important to know whether she was capable of laying at all or not.

The writer was amazed to find this state of things when, some years ago, he turned his attention from managing woolen-mill interests to trying to manage a poultry-yard. But, in spite of the fact that he was wholly unable to find bird or strain that were known to be exceptional egg-producers, he succeeded, within six years after starting, in building up a flock that averaged annually considerably over 200 eggs per hen.

Before deciding to publish this work, I found, after diligent inquiry among the leading poultrymen of the United States and Canada, and some correspondence reaching to other countries, that there was no known method—other than the slow and costly one of trap-nesting—of selecting birds of great egg-producing capacity. Trap-nesting, in addition to the faults mentioned, which makes it almost impracticable for the farmer, had a still more serious one in the writer's judgment; it could not trap-nest roosters, which I have found to be more than "half the flock." For this seemingly insurmountable difficulty I have found an easy solution, and can as readily identify the male as the female, and as unerringly.

The facts of which this document treat are a discovery, a method, and a development all in one. The happy inspiration and discovery came within a few hours; but it has reached this workable and absolutely reliable form by a costly analytical

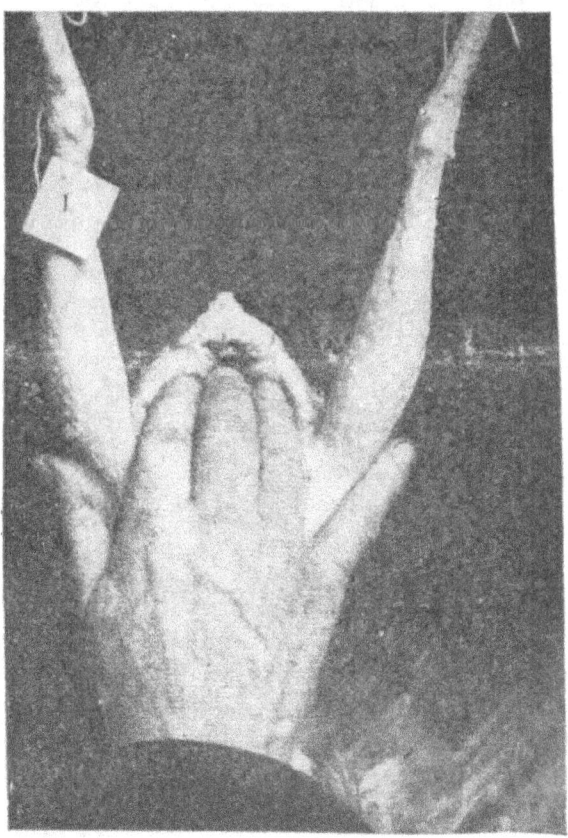

Cut No. 1—A Leghorn hen showing this development has the egg-laying instinct at its maximum.

126 THE CALL OF THE HEN.

and experimental process extending through years. After the underlying principle had been found, it had to be tested and proved to my own satisfaction. Then the various objections and criticisms, which will occur to many readers, had to be answered or met by actual practical experiences.

The method enables one:

1. To easily and without error weed out all the worthless birds from a flock; those that do not lay at all, also those that lay so little that it is a loss to keep them. This alone means millions to this country.

2. To separate just as unerringly all pullets before they begin to lay; indicating the coming great layers, the fair layers, the very poor, and the barren. The latter will be found in nearly all flocks.

3. To tell those not liable to lay when disposing of old or other hens for the table or market or for other reasons.

Beginning my investigation, as I was compelled to, with birds selected wholly without egg-record, I was soon greatly impressed with the dissimilarity of formation of the pelvic

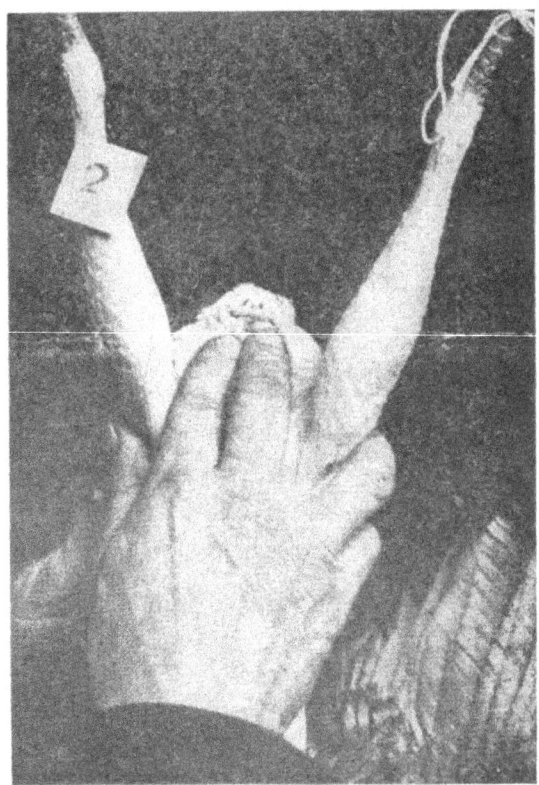

Cut No. 2—This is a hen of medium development. She is a fair layer.

bones and surrounding portions of the body, particularly of the former. Some I found nearly closed up, hard, and unyielding; others barely admitting one finger between these points; while a very few would easilly admit the ends of three fingers between the tips of the pelvic bones, and these were generally thin, tapering, and elastic. With this clue, I was not long in finding that my great layers were the latter and my barren and nearly barren ones the first mentioned. My attention was next forcibly called to this by seeing a long row of dressed pullets and hens in a butchering establishment. Noticing the great difference in the formation, I secured the privilege of numbering the hens and having the entrails, as they were removed, left by the side of each bird. In every instance I found my suspicion verified; the indications of large numbers of eggs and ample machinery to go with them, with the wide, pliable pelvic bones; and just the opposite condition with the narrow ones, the very least, or no egg indications whatever, with the bones very close

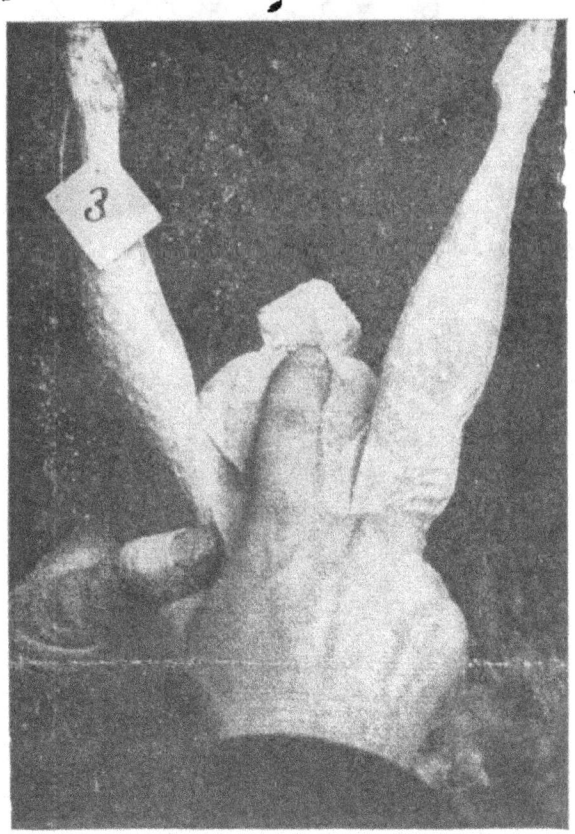

Cut No. 3—Hens with this development are of little or no value as layers.

together at the points and unyielding to pressure, hard, thick, and rounded in. This experiment was tried again and again, with different breeds, but never with different results.

I was satisfied I was on the right trail now, and determined to spare neither time nor money to make sure I was right. For several years following these discoveries I spent much time and money visiting well-known poultrymen and others, frequently paying as high as $10.00 for best known layers, only to kill them to prove or disprove my conclusions—to photograph the live bird, next her dressed body, then her skeleton. In every instance I found my theory corrert. I divided my own flock according to my findings into three flocks, and the very first day's lay proved my theory beyond question, so far as one day could. I then divided other and many flocks; but wherever they were and whatever breed, without an exception the same result followed.

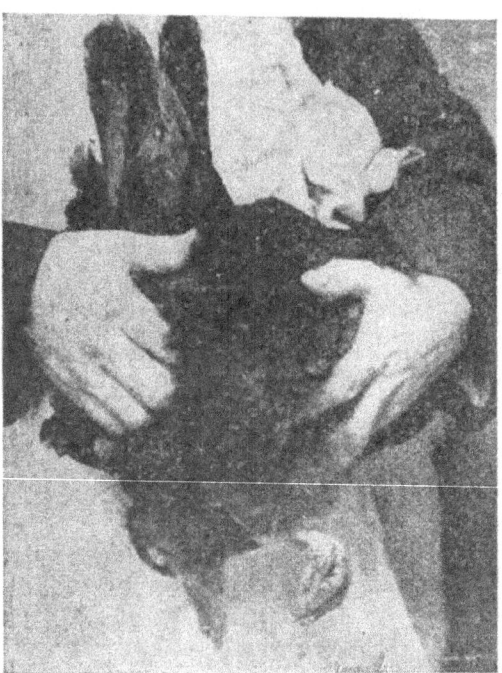

Cut No. 4—Showing a convenient method of holding fowls when testing them.

Skipping a number of years, I might say right here that in 1904 I divided the flock of Leghorns, Wyandottes, and Plymouth Rocks at the Minnesota Experiment Station at Crookston into three pens: first, the best; second, medium to poor; third, very poor or barren. I was about twenty-five minutes doing this in the presence of C. S. Greene, at that time the manager, whom nearly all the leading poultrymen know, and Mr. T. A. Hoverstad, then superintendent of the station. These gentlemen then had absolutely no faith in the method! not knowing anything about it; but were assured by me that if the barren pen laid an egg or either of the others failed to perform as I indicated, they were at liberty to publish the method and me to the world as a fraud. The first day showed pen No. 1, 45 eggs; pen No. 2, 20 eggs; pen No. 3, no eggs; and this continued, with slight variations, the entire period of the experiment, which lasted for weeks; though not a single egg appeared in the barren pen. The per cent of eggs to the 100 hens for the entire time was: First pen, 60 per cent per day; second pen, 37 per cent; third pen, nothing. But for lack of room I might give many more experiments and tests fully as startling as the above.

But to go on: Within a few years after selecting my first layers in this way, I had a flock the larger part of which was laying 200 eggs and above per year, individual layers greatly exceeding this.

Then came another discovery, fully as important as the first. I noticed that, though I hatched all my pullets from the best layers' eggs, some of them were exceedingly poor layers; now and then one of them barren. I studied upon this for a long time, spent more money, and killed many more birds. Then with another idea, which as suddenly as the first dawned upon me, I made for the slaughter-house once more. I soon had a row of forty or so dressed male birds this time laid out before me; and then at a glance I saw my long-sought solution. There was the same great difference in the pelvic formation found in the hens. I examined my roosters to find that half of them were absolutely worthless. Why do I say that the rooster "is MORE than half the flock?" Because later I found, as many know, that the female offspring take largely after the father and the male offspring after the mother. It is so with all animals, and almost always so in the human family. Had I used males of my own raising, I should have done better, but I had not. By the way, I found two high-priced and "high-scoring" birds used at the Crookston Station in 1904 absolutely without value, and Mr. Greene now agrees with me fully that they were, although he was at the time quite indignant when I pronounced his costly beauties worthless.

I may say here that, while I found one very good exhibition bird in this experiment station flock that was wholly worthless as a layer, I am pleased indeed to be able to state that one bird which had taken several prizes for markings, etc., I found to be a priceless layer. I never saw but one bird that came anywhere near being that hen's equal. I found one, however, with very poor markings that outranked any hen but her.

From this time on breeding hastened matters fully as much as selection, and I soon had—or rather, to be accurate, at the end of six years from my first start I had a FLOCK AVERAGING CLOSE AROUND 250 EGGS EACH PER YEAR; A FLOCK PAYING ME MORE THAN DOUBLE THE PROFIT MY FIRST FLOCK COULD. During the last few years of this period I again and again, for experimental purposes, mated excellent hens with narrow-pelvic-boned males, and every time a crop of pullets that varied greatly in egg-yield was the result. Again and again I bred wide-pelvic-boned males with narrow-boned females with the same results. But wide-pelvic-boned males with hens of the same formation (with the exception now and then at far-apart intervals, a freak) brought excellent layers. Occasionally a male bird failed to transmit well, but this I afterwards found was only when it was wholly lacking in masculine qualities, as denoted by the width and depth of head and back of neck, with other indications common to masculinity in all other animals. From this time I began mating wide-pelvic-boned males with my widest hens and a marked increase in the number of great layers was evident—in fact, the third year it was the great exception to find anything but first-class layers among the pullets.

Its Advantages.

The advantages of this method for one owning even a small flock of birds are so apparent that space need not be given to discuss it. To one having a large flock it means, must mean, a small fortune in additional profit, with no more labor or investment; to those engaged in selling eggs for hatching it is bound to mean everything in the near future. It would be simply suicidal for a farmer, or anyone depending upon the eggs of his flock for the profit, to be so unbusinesslike as to buy eggs for hatching from untested flocks. We do not believe it would be possible to find one who would do so, after knowing from experiment stations and otherwise that the method is unfailing.

Some of the advantages over trap-nesting have been stated; perhaps the strongest being that we cannot trap-nest roosters. In addition, I might call attention to the fact that trap-nesting a single bird must extend over the entire year to be at all accurate, and would take many times the amount of time it would require—by this method—to settle the laying possibilities of a thousand pullets. A little more time would settle the laying powers of a large mixed flock at mixed laying seasons, which might require two or at least three examinations a week or ten days apart.

Again, a worthless pullet can be found when she is from five to six months old and fatted and sold without having to keep her a full year in order to do it safely. Besides, handling hens almost always tends to disturb and discourage laying. Trap-nesting will, if persistently followed the entire year, give nearly the exact individual record, which is not material to one egg man in a thousand. It cannot be exact, however, as a shut-in and otherwise disturbed hen never does her best.

This method applies to other birds as well—turkeys, for instance. Last fall I bought two turkeys for experiment; one was SMALL, with LARGE egg-development; the other LARGE, with SMALL egg-development. The small bird has laid and hatched out two litters of fourteen each the present season, and has at this date laid twenty-three eggs towards a third litter. The large one laid and hatched fourteen eggs early in the season, and has shown no signs of laying since, but has taken on much more flesh than the laying turkey. This would, in addition to indicating laying turkeys, also show what to breed if large birds only are desired—as would nearly always be the case with turkeys.

The absolute surety of never killing a bird for market or home consumption that is laying, about to begin laying, or is liable to lay in the near future, is another decided advantage over the trap-nest, and one of the quickest available advantages fo the system.

Again, the process requires no investment in patent nests, leg-bands, or other fixings, which amount, in trap-nesting, to many times the first and only cost of this method. For accuracy in all the advantages claimed for this method, we will most gladly submit to a test with the greatest expert trap-nester that can be selected, if it can be so arranged that some high authority in poultry matters or some Government Experiment Station shall have charge of it. This unconditional offer we make to the world.

How to Select.

As the basic principle of this method of identifying capacity for egg-production is the width and relative condition of the pelvic bones and surrounding construction, it is obvious that exact measurements cannot be given, unless a distinct breed be designated. A Cochin lays a large egg, and is built accordingly; a Bantam lays a small egg, and its pelvic development in inches is correspondingly smaller. It would be manifestly misleading to apply the same measurements to the two birds.

While the ability to make this allowance will come to the operator quickly—almost intuitively after a very short experience—I have thought best to confine all my descriptions and measurements here to one breed of fowls only, the Leghorns, these being a medium-sized, representative bird, well scattered over the entire country. It will be easy from the measurements to work up or down, as the birds on hand may be larger or smaller. It is all a matter of comparison, and, all things being equal, the bird with the widest and most pliable pelvic bones will be the greatest layer, while the one with very narrow contracted pelvic formation will lay little, if at all. Behind the pelvic bones lies the egg machinery, and it will be found more abundant and roomy the wider the bones.

SELECTING PULLETS.

(Leghorns.)

Perhaps the best time to select layers for a flock is when the pullets are from four to six months old. If all are in a uniformly thrifty condition at this time, it is next to impossible to make a mistake. The best pullets at that age should show a width of about 2 inches, while the best matured laying hens should show a development of about $2\frac{1}{8}$ inches. (See cut No. 1.)

Pullets of Plymouth Rocks and their class should be selected about a month later and then show slightly larger, about $2\frac{1}{8}$ inches. The best Asiatic pullet, about $2\frac{1}{4}$ inches at seven or eight months old; the Leghorns being earlier maturers. At the end of six years of careful selecting and breeding I found my Leghorn pullets quite as wide and well matured at four months as my first ones were at five months.

Second-class Leghorn pullets from five to seven months old will show a development of about $1\frac{5}{8}$ inches. (See cut No. 2.)

At six months old all Leghorn pullets showing only 1 inch or less pelvic development should be discarded, regardless of feather or comb. They will never make layers. (See cut No. 3.)

All things being equal, the earlier a pullet begins to lay the better and longer will she lay.

SELECTING MATURE LAYERS.

The next best time to ascertain a hen's laying qualities is when the whole flock is in the flush of laying—in other words, when about all are at work. Those found then to measure about $2\frac{1}{8}$ inches are extremely good layers. Some flocks have very few of these priceless birds in them, while others have good numbers. From this class of layers, and above that measurement, and from these only, should eggs be saved for hatching.

Occasionally hens are found measuring as high as $2\frac{3}{4}$ inches; these hens, with the best of care, will lay as high as 280 eggs per year; those measuring about $2\frac{3}{8}$ inches may be depended upon to go as high as 250. The fact that this kind of hen can be found is ample proof that with proper selection they can be bred in large numbers.

Hens found at this time measuring from $1\frac{7}{8}$ to 2 inches are real good layers, and should not be discarded, if one wishes to build up a large flock, but they should not be bred from. Hens in the flush of laying measuring only $1\frac{1}{4}$ to $1\frac{1}{2}$ inches are poor, and those showing from an inch down should be discarded, regardless of shape or color.

A large enough flock of the first-mentioned hens would make any poor man rich; the second kind would keep themselves and their owners going; while many of the last-named class would make a rich man poor.

Poor layers, kept well and fed a large variety of scraps and other foods, will sometimes make pretty fair egg records for a short time, and birds of the best quality, under exceptionally hard conditions, will make poor records. There are also occasional freaks in both extremes of measurements, but they are so infrequent as not to be at all important. Approximately, 280-egg hens that measure as high as $2\frac{5}{8}$ inches in the flush of laying will show about $\frac{3}{8}$ to $\frac{1}{2}$ inch less when not laying, and this shrinkage in measurement will apply to all other grades in about this proportion.

Selecting for Fall Marketing.

We do not like to kill birds about to begin laying, that are laying, or really good ones that are just through laying, particularly when there are plenty in the flock that do not come under any of these heads.

In this alone the cost of this method, when once well understood, can be saved several times in a single season with a good-sized flock of birds.

While the exceptionally-good layers can be told readily and at almost any time, laying or not, and an absolutely worthless bird can be told the same way, there is a time, just when the real good layer is resting and the common to poor layer is doing her best, when they come—for a short time only—close together in pelvic appearance.

While it is not safe to kill a bird that measures $1\frac{1}{8}$ inches or over, it is possible for a very fair layer to not be much wider than that at the close of laying out her litter. Some good layers, that in the flush of laying will measure $1\frac{3}{4}$ to 2 inches, at the close of their laying period will sometimes close up to about $1\frac{1}{8}$ inches. A very poor layer in the flush of her laying time might be $1\frac{1}{4}$ to $1\frac{1}{8}$ inches, so care must be taken at this period not to confound the two conditions, which do not exist at any other time. This is referred to in the Introduction. To wholly prevent this—when it is desired to save every at all good layer—it is well to make two or possibly three examinations, a week or so apart. In this way there will be no danger of confounding the one about to begin laying with the one about to quit, and the poor layer can be told from the good one.

When killing a whole flock at two or three years old, as many do, no hen measuring $1\frac{1}{8}$ inches and under is worth keeping; particularly is this true if the birds have been well fed and stimulated to about their full capacity. No hen of any value for egg-production will have an egg in her at this time and measure so small unless she is a slow, infrequent layer at her best. Sometimes this kind of a hen with very small measurements will be found laying an occasional egg late in the season.

Selecting Roosters.

We have said how important it is to have males of the right formation to mate with the great layers for breeding purposes; we need not emphasize this; it is so evident that we cannot trap-nest a rooster, and equally so that years of trap-nesting hens can be ruinously upset in a day by crossing with an inferior male, that it would reflect upon our estimation of the reader's intelligence to say more about it.

I have found Leghorn roosters that measured $1\frac{3}{4}$ inches, but they are rare and priceless. A good matured bird should measure $1\frac{1}{8}$ inches and a pretty fair one 1 inch. I would not use one that measured less, if I could possibly help it. Many fine-looking birds measure only $\frac{1}{2}$ inch, but such ones will ruin the offspring of the best layers and should be discarded, whatever their qualities in feather, tip of comb, or anything else.

Now and then the objection reaches us that the high-type roosters referred to cannot be found. I have found them, as others have, and I believe there are nearly or quite as many in proportion as there are of the 250 and above hens; but we do not save all the roosters as we do all the pullets, and they are correspondingly scarce among mature males. By selecting always from large numbers of males before they are killed off this objection will be largely and quickly overcome.

The fact that males of this class can be selected is of itself a discovery sufficient to revolutionize the whole poultry business without the examination of a single hen—were time enough taken; but the two together bring absolute and immediate results.

In the hands of a slightly experienced or an at all competent person the element of chance is entirely removed by this method of selecting layers and males; and one is just as sure of the results sought as that a hen will die if her head is cut off.

We ask but one thing: that judgment be withheld till this method be tried. If the tests are fairly conducted, there can be no failure.

Crude infringements and imitations of this discovery and System—as of everything else of value that has cost years of investigating and experimenting—are liable to spring up, but the safety and economy of going direct to the fountain-head need scarcely be suggested.

Dated November 20, 1904.

THE END.

www.ingramcontent.com/pod-product-compliance
Lightning Source LLC
Chambersburg PA
CBHW081118240526
45470CB00019B/2489